Linux PC-based Measurement Electronics

hardware & software

Yury Magda

Linux PC-based Measurement Electronics

hardware & software

Yury Magda

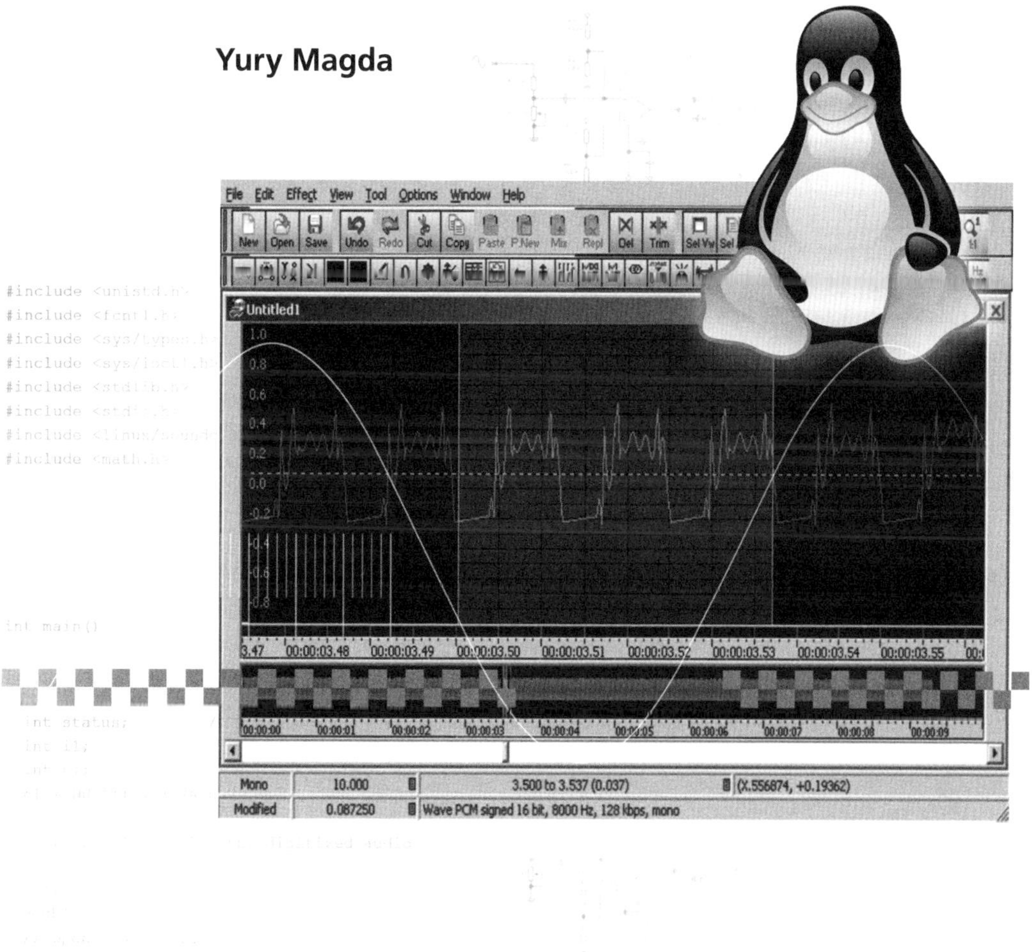

British Library Cataloguing in Publication Data
A catalogue record for this book is available from the British Library

ISBN 978-1-907920-03-5

Prepress production: Jack Jamar | Graphic Design, Maastricht
First published in the United Kingdom 2011
Printed in the Netherlands by Wilco, Amersfoort

Second Edition, February 2012

109038-1/UK

Contents

Preface

Nowadays personal computers have become an integral part of our life. You probably already realized that PC technology (known as 'embedded PCs') has been adopted into many systems intended for automation and control in laboratories and industry. In possessing the advanced computational capabilities of a modern personal computer, and under the control of a lean, fast operating system such as Linux, those systems can readily assimilate measurements from complex physical quantities, control equipment or provide complex data acquisition and analysis. PC technology extends still further, into communication systems, home automation and so on.

This book is written for both hardware and software hobbyists and developers who would like to learn how to use a personal computer as a control centre in embedded projects; thus, designing and building various measurement and control systems to be controlled by their desktop or laptop PC running a Linux operating system. The PC's present, basic peripherals (serial/parallel ports and sound card) are actually already well suited for many measurement and control systems. PC hardware resources provide a convenient interface for appliances to measure temperature, humidity, pressure, light or other physical quantities. Additionally, PC peripherals can be used to trigger motors, relays, lamps and other loads depending on some predetermined conditions. By combining specific input and output signals travelling through PC peripheral devices a developer can quickly build embedded systems implementing even complex tasks.

Designing embedded systems usually involves both hardware and software. As the electronics and programming are often regarded as separate areas of engineering that almost never cross, you might otherwise assume that working with both hardware and software is an extremely difficult task. This book makes the task quite manageable.

Linux has been chosen as the working operating environment for all of the projects. You will need to be already relatively comfortable or, at least, keenly enquiring in a Linux environment and, as such, must be fairly familiar with basic shell commands.

The software part of the embedded projects described herein is written in C. You should, ideally, have some knowledge of programming in this language. On the other hand, you don't need to be a guru in C programming - all programs from this book are very simple and clear, so keen beginners will easily understand the source code. I also assume that you have at least a rudimentary understanding of local area networks and the Internet - which will help in your understanding of how some of the distributed applications work.

Most embedded projects described in this book deal with some straightforward electronic circuits connected to the PC. Those circuits must be built manually, so some experience in assembling and testing simple electronic devices will be useful. Luckily, assembling and testing of the circuits presented here will not require an in-depth knowledge of electronics or module functionality - you only need to interface the prescribed circuitry to the PC. However, if you wish to develop circuits further for yourself, the relevant timing diagrams are described in detail. All of my external circuits have been built and tested using low-cost and widely used electronic components; allowing you to design simple, yet effective electronic devices.

You will see that it really is possible to build your own complete PC-based embedded systems by using simple low cost (or zero cost) software tools and low-cost electronic chips. I hope that this book encourages you in your own future designs.

Acknowledgments

I'm indebted to many people who helped make this book a reality. I'd like to thank all of the engineers and developers who gave their time to review and evaluate the material in the volume. Thanks to Paul Ridgway who once again did an exceptional job of reviewing and editing my author's copy. I really appreciate all the help -thank you very much. I'd also like to thank all of the folks at Elektor who did a lot of good work to help put this book together. Finally, I wish to thank my wife, Julia, who's always there when the going gets tough.

Chapter 1. PC in home projects

Nowadays the personal computer is readily recognised as a powerful design tool in office, scientific and industrial applications. The modern desktop PC or laptop may also serve as a basis for developing various embedded systems for home and laboratory experiments. With our PC we can measure physical quantities, control actuators and build home security systems, to name but a few. Often such "Do-It-Yourself" (DIY) systems can compete with industrial applications in terms of performance and capabilities. The other advantage of DIY projects is that they can be very inexpensive.

By combining innovative approaches you can even create your own, powerful, PC-based home research center or configure your PC as a home control center. You can also design your PC-based home system to be controllable remotely using Internet or e-mail. Nowadays, modern technologies help individuals to build very complicated systems because many solutions that were too expensive a while ago have become cheaper and accessible for the professional developers and hobbyists – for instance, wireless technologies. Another aspect that allows construction of complex systems by individuals is that modern chips are usually multi functional and programmable; this significantly simplifies design of embedded systems.

Manufacturers often provide the familiar, standard interfaces (SPI, I^2C, etc.) for configuring their chips, so a developer does not necessarily need to know the inner, sometimes very complicated architecture of a particular chip – it's enough to configure some hardware registers and start working. This book is dedicated to designing home and laboratory systems running on the PC with a Linux operating system. All the projects discussed here will work on a desktop PC or laptop with parallel and serial interfaces built-in or installed. Some projects require a sound card, for which you can use either a built-in device that resides on a motherboard or take a stand-alone sound card.
Using Linux for these purposes has some great advantages over Windows. The most obvious advantage of using Linux is that it is free. Also, with Linux, you have the power to control just about every aspect of the operating system.
As we want to build embedded projects we surely must control hardware. This task is essentially simpler in Linux than in Windows because the Linux operating system provides easy access to hardware registers directly from user software written in C. Linux is perfect for old computers which can be easily transformed into home laboratories and control centers. That is almost impossible with Windows because of its great demands concerning hardware resources. Compared to Windows, using Linux requires knowledge of the operating system, patience and a strong learning curve. You must have the desire to read and figure things out on your own, rather than trusting everything to the OS. To start with, we need a Linux operating system to be installed on a computer. All the projects designed for this book have been tested under Ubuntu 10.10 desktop version of Linux, but you can

use your own favourite version of Linux for design as well.
Desktop Ubuntu is a very popular light-weight version that exhibits good performance and provides a user-friendly GUI interface. Generally, you don't need to install a full system on a hard disk if you don't want to. Instead you can acquire a LiveCD (single CD bootable version) of that operating system and work exclusively from that CD! The other popular Linux distribution which you can use for building embedded systems is Knoppix (last version is 6.4.4). Knoppix may also be loaded from a LiveCD if needed.
Designing embedded projects from this book requires some basic skills in electronics and programming. The software source code was designed using the C language, so you should be familiar at least with the basics of programming in this language. As our projects also include hardware, you must know how to assemble and test your circuitry.
All embedded projects from this book may be represented in general as shown in Fig.1.1:

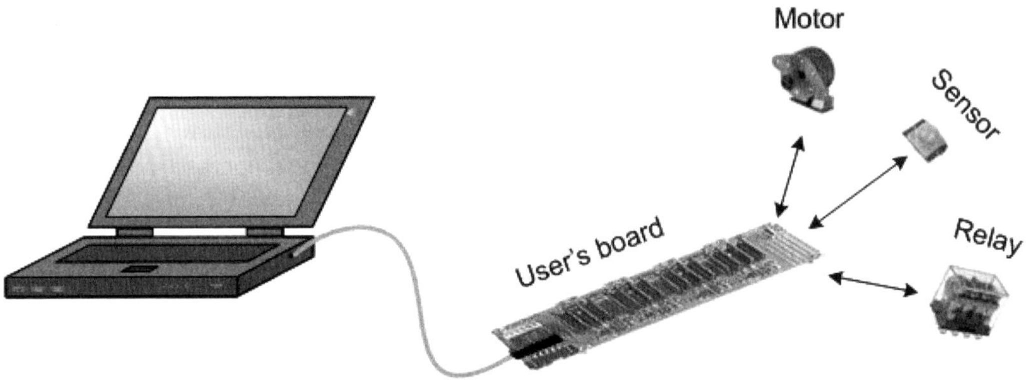

Fig.1.1

As you can see, our system may capture signals from some sensors, analyze data obtained and issue control signals for actuators (motors, relays, heaters, etc.). All the signals from sensors pass through some external circuitry (called the "User's board") to the PC where they can be processed by control software. The control program running on the PC executes some algorithm which analyzes data received and makes "decisions" about a behaviour of our system. In the simplest case the "User's board" provides some kind of hardware interface (SPI, I^2C, etc.) between sensors/actuators and a PC and does nothing else.

On the other hand, our "User's board" may be designed with some "intelligence" to make a rough data processing prior to passing data to a PC. In this case such a "User's board" could be equipped with a micro-controller or microprocessor. Most of our projects don't require a separate "intellectual" external module so we use very simple development boards without any "intelligence". Nevertheless, we will discuss a couple of the projects with micro-controller boards when designing distributed systems.

When we talk about 'PC-based projects' we assume that we don't need to deal with "real-time" high-speed events. What does this mean? Suppose for instance, that we want to build a system where we need to read data from some sensor every 1 microsecond. If we try to realize such a system under Linux or Windows, we will more than likely encounter serious problems because of a 'tight timing'. In this case a proper solution requires using either real-time operating system installed on a PC or an "intellectual" development board with a microcontroller/microprocessor. Luckily, many real-life systems don't need to have such performance, so we can freely use control software developed under Linux in our embedded projects.
Many times you need to control your embedded system remotely, for example, from your workplace or when you are or when you are travelling. This comes in handy when you have to get data from a remote system implementing some continuous task or controlling some process by switching on/off actuators. Modern technologies allow you to accomplish remote control using wireless communications and/or the Internet. The very general scheme of such a distributed system is shown in Fig.1.2:

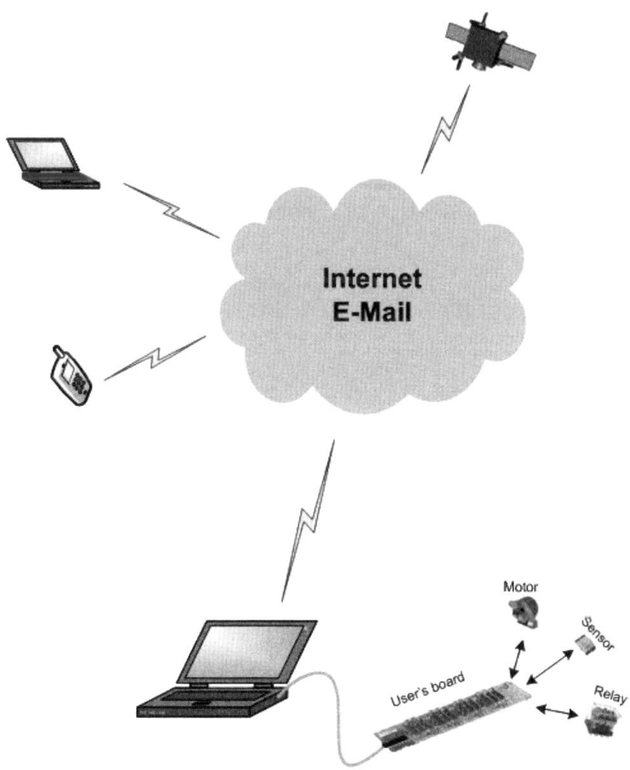

Fig.1.2

Every project in this book, developed for use on a local PC, can be easily fitted into a distributed system comprising different computers scattered all over the world. We can select various methods for designing such distributed systems, although the easiest way is to use TCP/IP network technologies or e-mail. Both approaches will be discussed in Chapter 5 "Network and wireless systems for home and laboratory".

PC peripherals in embedded projects

Embedded projects described in this book are built using parallel/serial ports and a sound card. Here we discuss in brief each of these interfaces. The parallel port (or parallel interface), being one of the oldest interfaces of the PC, is still frequently used in embedded systems for interfacing to external circuitry. The standard parallel interface allows you to output (write) data via 8 lines and to input (read) data via 5 lines, although we can easily expand the number of lines by adding some circuitry.

The parallel port of the PC usually resides on the motherboard for desktop PCs and may be accessed via three hardware registers (ports). It's possible to add extra parallel interfaces to the PC by using one of the numerous expansion boards available. As for laptops, you can easily select numerous PCMCIA-to-parallel or USB-to-parallel expansion cards to add. Attaching external circuitry to the parallel port is implemented via a DB25 standard connector on the back panel of the PC for a built-in parallel interface or via the same connector on the expansion card.
Embedded projects where a parallel interface is used are discussed in Chapter 2 "Parallel port in the home design".

There will be discussed, for instance, a simple 12-bit digitizing system, a wide-band oscillator working from 1Khz through 68MHz, digital-to-analog converters with digital potentiometers and other projects.
Another popular hardware resource is a serial port (or serial interface). This interface is based on the RS-232 standard and is widely used in designing PC-based measurement and control systems. The serial interface comes in handy when we want to build a low-cost reliable data transfer systems with minimal effort. The serial port also serves as a basis for RS-422/RS-485 protocols which are used for building distributed systems. With new high-performance UART chips it's possible to work at 921K Baud.

Many modern PCs are not equipped with the serial port interface, so we need to acquire one from the numerous expansion boards containing two or more serial ports.
The heart of the serial port is the UART (Universal Asynchronous Receiver/Transmitter) that provides all the signals required for the RS-232 interface. The UART is completely controlled via hardware registers of the PC, so we can configure this interface and triggering signals. The PC's serial interface provides standard RS-232 levels of signals that are not compatible with single supply

analog and digital circuits, so we need to use some RS-232-to-TTL converter which, in our projects, will be realized by using an industry standard driver/receiver chip from Maxim Inc., the MAX232. We will consider various projects including digitizing systems for analog sensors, systems that controls temperature and humidity, etc. All this will be discussed in detail in Chapter 3 "Home projects with a serial port".
Last in the list of peripherals which we want to discuss is a sound card. Usually the sound card comes as a built-in device on the motherboard, although you can still acquire stand-alone modules for various experiments. We usually apply a sound card in embedded projects when we need to generate various signals going to external circuits and/or capturing signals coming from external sources for further processing. There are a lot of applications called "software oscilloscopes" that implement both these tasks, although most sound cards can reliably process relatively low frequencies, usually 20–40 KHz. However, such oscilloscopes come in handy as a cheap alternative to professional devices, especially when we need to measure low frequency signals. Programming sound cards in Linux is a complex task enough, so we will discuss how to design our home sound laboratory using free software and a popular scientific application, Matlab/Simulink by The MathWorks, Inc. The latter is not completely free, though you can use a low-cost Student Edition for building your home or laboratory applications. We will discuss using a sound card in embedded projects in Chapter 4 "Sound card in home projects".

Software development

All the projects from this book require some control software. The control applications used were developed. Developed using a standard and free `gcc` compiler, which stands for the "GNU C Compiler". The latest version of `gcc` applied to the development was 4.3.4, though earlier versions can be used as well. Writing applications in `gcc` requires some basic skills in C programming, so I expect that you are familiar at least with the basics of C.

The C language has some great advantages over other popular software tools, such as Java, Python, etc. Languages like Java and Python shield the programmer from having to worry about memory allocation and pointers. That is usually a good thing, since operations with memory are error-prone when building a high–level program. On the other hand, when dealing with low level code such as the part of the operating system that controls a device, C provides a uniform, clean interface by using a group of special functions (we will discuss this later in this chapter).
The Java, Python and PHP languages are widely used for writing high-level applications having graphical user interfaces (GUI) having numerous extended capabilities, but all these tools are inappropriate for low–level operations.

Implementing the control over hardware within, say, a Java or Python application is essentially a complex task and requires using special external modules (for example, pySerial or pyParallel in Python) being developed for accessing hardware registers. That doesn't simplify the task as much as it might seem because of performance and compatibility problems. Another frustration when using such modules is that some features, such as accessing bits in hardware registers (a common task when dealing with hardware) or even reading some registers are poorly supported – that induces a developer to write a complex and, sometimes, ugly code for low–level processing.

As a result, the performance of low–level I/O operations with hardware degrades causing a loss of data accuracy, especially when you write a control code for time–dependent processes such as an analog–to–digital or a digital–to–analog conversion.

In these cases C code comes in handy. You may put all low–level control code written in C into a shared library, where functions may be easily called from Java, Python, etc by using standard methods defined in these languages. This approach works forever and everywhere regardless of Linux distributions and Python/Java/PHP/etc. releases. Using "pure" C for implementing low–level I/O operations over hardware saves a lot of time and effort. From this point of view I have simplified the C source code so that it could be possible to easily build a shared library, if required. Of course, the C source code may be compiled as a usual console application as well.

Another approach which we will discuss in Chapter 5 is based on using 'sockets'. We can write, for example, two standalone applications in different languages, say, C and Python, which can interact with each other using sockets. For example, we can write in C a simple daemon that will acquire data from some sensor and pass this data through a local or TCP socket to another application written in Python. In this case we don't need to adapt the Python application for implementing operations with hardware – all that we should do is to provide code for reading data through the socket dedicated for both applications.
Naturally, this approach requires knowledge of network programming in C and Python. The same is true for Java or PHP.
The embedded projects we will choose to consider herein require accessing of PC peripherals via hardware registers (ports), so we need to apply some library functions to accomplish this task. These functions are specific and exist only in Linux systems developed for the x86 platform. We can use abbreviations `in*` and `out*` when describing these functions. The `out*` functions perform output to the port, while the `in*` functions perform the input from the port. Alternatively, it's said that `out*` functions write data to the port and `inp*` functions read data from the port.

The asterisk (*) denotes the size of data processed by the `out*` and `in*` functions. For example, the b-suffix functions (`inb()`, `outb()`) are byte-width and the w-suffix functions (`inw()`, `outw()`) are word-width, etc. As to our projects, we deal only with a single byte, so we apply `inb()` and `outb()` functions. The above functions were primarily designed for internal kernel use, but can be used from a user application. This may require **-O** or **-O2** options to be included for compiling source code, although the later versions of the `gcc` compiler (included in Ubuntu, Knoppix and other modern versions) allow excluding these options.

Before a first call of the `inb()` or `outb()` function wherever inside the program, one of the `ioperm()` or `iopl()` functions must be executed. Both these functions tell the kernel to allow the user's application to access the I/O ports – this gives port access permissions to non-privileged tasks. Failure to do this will cause the application to receive a "Segmentation fault" error. As we will use these functions very frequently, let's look at them in detail.
First of all, the program where `ioperm()` and/or `iopl()` functions are called requires the root privileges when started. In order to provide root privileges for a given application we need to type a `sudo` command before the name of the program to be executed. For example, if we try to execute the application named `myprog` we have to type:

```
$ sudo ./myprog
```

Some systems restrict you in using the sudo command in some way, in which case use a `su` command instead of `sudo` to become a superuser (`root`) temporarily and launch the program. Another way to execute the program that talks to hardware is to login as `root`, but in this case you should be very careful while working in the operating system with such privileges!
You should also remember that both functions work only with the x86 architecture; other architectures don't support these given functions – they don't exist or will always return an error.
The `ioperm()` function has the following synopsis:

```
int ioperm(unsigned long from, unsigned long num, int turn_on);
```

As you can see, the function takes 3 parameters. The `from` parameter points to the starting port address which is affected by the `turn_on` parameter. The `num` parameter holds the number of bytes which will be affected starting from the port address `from`. Setting `turn_on` into "1" allows read/write operations, while clearing this parameter forbids access to the given ports. Using the `ioperm()` function requires `unistd.h` (for libc5) or `sys/io.h` (for glibc) header files to be inserted at the beginning of a source code. If you are not certain about which file to use you can simply include both in the source code.
The `ioperm()` function affects only the first 0x3ff I/O ports – this is enough to get access to the standard parallel port with addresses beginning from 0x378. But if we have some expansion board with a built-in parallel interface, the hardware addresses of this parallel port may be outside of the range of 0x3ff. In this case we need to call the `iopl()` function that allows you to access the full I/O space which spans 65536 hardware ports. Permissions the `ioperm()` function grants are not inherited for a new process created by a `fork()` function, although this takes place when an `exec()` function is called.
The `iopl()` function allows you to change I/O privilege level for a given process and has the following synopsis:

```
#include <sys/io.h>
int iopl(int level);
```

The I/O privilege level is held in a single parameter `level`. This function affects all 65536 addresses allocated for the system I/O space. The I/O privilege level for a normal ("user") process is 0, while the highest level may be equal to 3 – the value we will use in our applications. This function also affects the processes created by one of the `fork()` or `exec()` calls.
All functions listed above are used in all projects concerning parallel and serial interfaces of the PC. This is all about programming tips concerning the development of our control software. Besides writing C applications we will also operate with a Matlab/Simulink development system which we use for a development of some projects based on use of the serial port and sound card. This is a very powerful tool that allows building fantastic systems where you can combine signal processing and scientific calculations. We will discuss some useful projects in MATLAB and see how to build our own research centre in Linux. You can purchase the student edition of MATLAB directly on www.mathworks.com – it costs only $99. Note that a student email address is required to activate the product. The product is also available on www.academicsuperstore.com and www.gradware.com.

Chapter 2. Parallel port in the home design

A parallel port is very convenient for building various embedded systems to be controlled from a PC. In this chapter we will discuss how to operate with the parallel port's hardware and design some embedded projects which may be useful at home or in a laboratory. We will discuss the electronics we need and software which controls our circuits, but first we should recall how the standard parallel port hardware functions.

Parallel port hardware

When you begin to plan your projects you usually have two alternatives: use a standard parallel port of the desktop PC located on the motherboard or apply some stand-alone expansion board with a parallel interface built-in. Nowadays some desktop PCs and most laptops/ netbooks are not equipped with a parallel interface, so you would need to obtain one of the numerous low-cost expansion boards for your embedded projects. All projects discussed in this chapter were designed using the PC's parallel interface, but they are valid for any external parallel interface. The only thing you should need to do when using an external parallel interface is to alter the hardware addresses in the source code of the control programs.

From a developer's point of view, a parallel port offers 8 lines to output (write) data to external circuits and 5 input lines to read data from external devices. Control programs may interact with a parallel port via three hardware registers (called "ports") that accept reading and writing operations. External circuitry which we will design is attached to the parallel port via a DB–25 standard connector on the back panel of the PC or via the same connector on an expansion board.

The parallel port hardware provides 4 control lines, 5 status lines and 8 data lines. All signal lines can be accessed via three hardware registers called "data register", "status register" and "control register". These registers are frequently referred to as "data port", "status port" and "control port", respectively. Fig.2.1 shows the physical layout of the bits of the hardware port and their respective DB–25 pins:

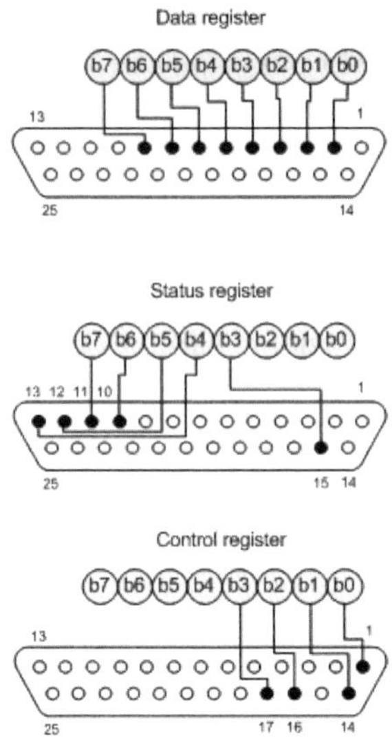

Fig.2.1

Below in Table 1 you can see a short description of each bit of the parallel interface.

Table1. Bit assignments of the parallel interface

DB-25 pin	Signal	Direction	Register/bit	Inversion
1	nStrobe	Out	Control/0	Yes
2	Data0	In/Out	Data/0	No
3	Data1	In/Out	Data/1	No
4	Data2	In/Out	Data/2	No
5	Data3	In/Out	Data/3	No
6	Data4	In/Out	Data/4	No
7	Data5	In/Out	Data/5	No
8	Data6	In/Out	Data/6	No
9	Data7	In/Out	Data/7	No
10	nAck	In	Status/6	No
11	Busy	In	Status/7	Yes
12	Paper-Out	In	Status/5	No
13	Select	In	Status/4	No
14	Linefeed	Out	Control/1	Yes
15	nError	In	Status/3	No
16	nInitialize	Out	Control/2	No
17	nSelect-Printer	Out	Control/3	Yes
18-25	Ground			

Before using a parallel interface we must know its representation in the Linux operating system. To identify the parallel interface(s) in Linux we must type the dmesg command in the terminal window. Below you can see the output that the above command produced on my PC:

```
$ dmesg | grep parport
[   12.099395] parport_pc 00:08: reported by Plug and Play ACPI
[   12.099456] parport0: PC-style at 0x378, irq 7 [PCSPP,TRISTATE]
[   12.330428] parport_serial 0000:01:05.0: PCI INT A -> GSI 21
(level, low) -> IRQ 21
[   12.330458] parport1: PC-style at 0x9800 [PCSPP,TRISTATE,EPP]
[   13.089058] lp0: using parport0 (interrupt-driven).
[   13.090917] lp1: using parport1 (polling).
```

In the Linux operating system, parallel interfaces are denoted as `parportX`, where X = 0, 1, … From the above listing you can see that the given system recognized two parallel ports which are `parport0` and `parport1`. The `parport0` interface is usually assigned a base address 0x378 (hexadecimal) that is the address of the data port. Next address that is "base address + 1" belongs to the status port (0x379); base address + 2 that is 0x37A is assigned to the control port. In all our embedded projects we will apply 0x378 and, if needed, 0x379 addresses that are the most common values for desktop PCs.

The second parallel port that is denoted as parport1 resides on the PCI expansion board and has the base address 0x9800 assigned by the PnP subsystem. Both parallel interfaces can also be controlled via separate interrupt lines denoted as "irq" (irq 7 is assigned for `parport0` and irq 21 for `parport1`). We don't use interrupts in our projects for this requires writing a device driver that is beyond the scope of this book.

Sometimes you could encounter the parallel interface with a base address 0x278 instead of 0x378. As we use 0x378 and 0x379 addresses in our projects, you should change these addresses on proper values everywhere in the source codes. If, for instance, I wanted to use the expansion board with the base address 0x9800, I would change 0x378-0x379 to 0x9800-0x9801.

The data port may be written to/ read from, the status register may be read and the control register may be read from/written to. The standard parallel port works with TTL-compatible levels of signals, therefore we can build our circuits using ordinary analog and digital chips that need a single +5V power supply. The current, the interface can sink and source, may differ from port to port. Usually, a value of current may be about 10-12 mA, so external circuitry should be attached to the parallel port via buffers.

We can read an external digital signal through status port and output a digital signal through data port. Although we can still read the data port, we haven't got the benefit of this because the byte read represents a state of output latches of the data port, nor the true input signal.

Let's see how one can operate the parallel interface under Linux by examining a simple project which toggles the LED attached to some bit of the data port. This project mainly demonstrates a common programming technique we use when dealing with the parallel interface.

Switching a LED

In this project we will see how to access, read and write parallel port hardware using `ioperm()`, `inb()` and `outb()` functions described in the previous chapter. Our application will toggle the byte of the data port (base address 0x378) every 10 seconds.
Open a text editor and type the following source code (Listing 2.1):

Listing 2.1

```
#include <stdio.h>
#include <sys/io.h>
#include <unistd.h>

int main(void)
{
  int bRead;
  ioperm(0x378, 2, 1);
  while(1)
    {
      bRead = inb(0x378);
      bRead = ~bRead;
      outb(bRead, 0x378);
      sleep(10);
    }
  return 0;
}
```

After you have typed in that source code save it, for example, in a file called ioperm_example.c. Recall, that we want to access hardware, so we inserted a statement with the `ioperm()` function before we call any of `inb()/outb()` functions. We use a data port of the standard parallel interface, so the first parameter of the `ioperm()` function will be the address 0x378. As we try to access the first two ports, the second parameter of this function is assigned value of 2. Thus, after executing the statement

```
ioperm(0x378, 2, 1);
```

the application has got access to the ports 0x378 0x379. We could set this parameter to 1 since we need to operate with port 0x378 only. The third parameter of `ioperm()` is set to 1 that allows reading/writing operations with those registers. Setting this parameter to 0 forbids an access to these ports from the user application.

Switching on/off of data port latches occurs in the endless `while()` loop. Each iteration begins by reading the byte from the data port latches into the `bRead` variable by calling the `inb()` function. This function takes a single parameter that is the address of the hardware port (0x378, in our case). After a byte has been read the program reverses all the bits of the `bRead` variable. The inverted value of `bRead` is then written back to the data port by executing the `outb()` function which takes two parameters. The first parameter contains the value of the byte to be written, while the second points to the address of the hardware register that, in our case, is 0x378.
The `sleep()` function ends the current iteration implementing a 10 s delay before the next iteration begins. The compilation of the source code can be accomplished by typing the following command in the terminal window:

```
$ gcc -o ioperm_example ioperm_example.c
```

After successful compilation we can execute the application by entering

```
$ sudo ./ioperm_example
```

To break execution of an application we can press Ctrl+C.
To test our application we can attach a multimeter to the parallel port. One probe of the multimeter should be tied to the pin 24 or 25 ("ground" of the parallel interface) of the DB–25 connector, while the other probe touches any of the pins 2–9. While the application is running you can observe the voltage swing on the selected pin every 10s. Another way to test our application is to assemble a simplest circuit with a LED and resistor attached in series to one of pins 2-9 as shown in Fig.2.2 (here pin 2 is used):

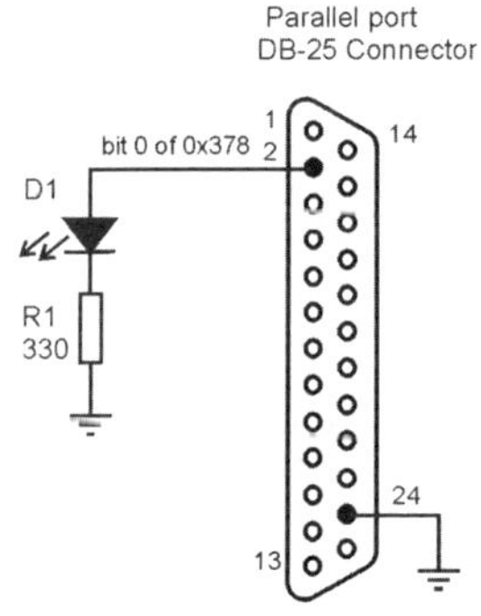

Fig.2.2

The following source code represents using the `iopl()` function instead of `ioperm()` (Listing 2.2):

Listing 2.2

```
#include <stdio.h>
#include <sys/io.h>
#include <unistd.h>

int main(void)
{

  int bRead;
  iopl(3);

  while(1)
    {
       bRead = inb(0x378);
       bRead = ~bRead;
       outb(bRead, 0x378);
       sleep(10);
    }
 return 0;
}
```

As you can see, the `iopl()` function takes only a single parameter - assigned a value of 3 (the highest level of privileges). After saving the source code in the file iopl_example.c we can use the following commands to compile the source code and launch the application:

```
$ gcc -o iopl_example iopl_example.c
$ sudo ./iopl_example
```

To become more familiar with the parallel port let's consider one more example. Here we will toggle bit 0 of the data port (address 0x378) and read the value of this bit via bit 5 of the status port (address 0x379). See the simple schematic circuit for this experiment (Fig.2.3):

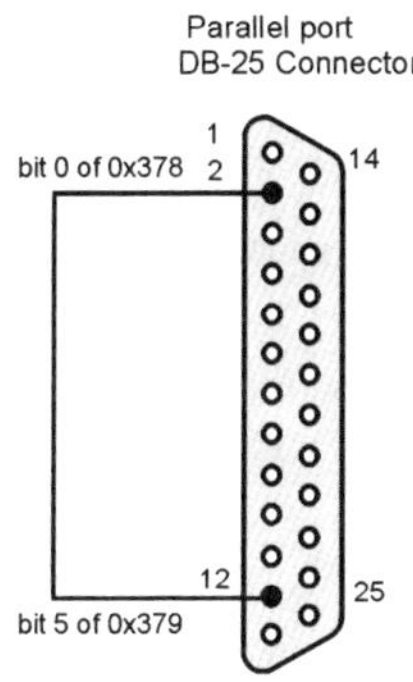

Fig.2.3

Below you can see the source code of this application which toggles bit 0 of the data port every 10 s (Listing 2.3):

Listing 2.3

```
#include <stdio.h>
#include <sys/io.h>
#include <unistd.h>

int main(void)
{
 int bRead, bWrite = 0x0;
 iopl(3);
 while(1)
   {
     outb(bWrite, 0x378);

     bRead = inb(0x379) & 0x20;
     if (bRead != 0)
     printf("Bit 5 is set.\n");
   else
     printf("Bit 5 is clear.\n");
   bWrite = ~bWrite;
   sleep(10);
 }
 return 0;
}
```

The `bWrite` variable holds the byte which is written to the data port; the value of `bWrite` is reversed after the byte has been written to the data port. Each iteration repeats every 10 seconds (function `sleep(10)`).
The `bRead` variable holds the value of bit 5 of the status port. As pin 2 and pin 12 are tied together, the value of bit 5 echos what has been written to bit 0 of the data port. The program code separates bit 5 of the status port by executing the following statement

```
bRead = inb(0x379) & 0x20;
```

Depending on the value of the `bRead` variable the program outputs one of two messages by calling the `printf()` library functions.
By now we know, in general, how to operate with the parallel port programmatically, so we can start designing more useful and complicated projects. The following project will be a simple thermostat system that switches on/off a heater when some temperature conditions are met.

Simple thermostat system

This demo project shows a practical small thermostat system controlled via the parallel port. Suppose we want to build a system that could control the temperature inside an oven. Our system may function in various ways, but we select the simplest algorithm described below.
When a temperature exceeds a given threshold a heater element is turned off. Otherwise, when the temperature inside the oven drops below the same threshold the heater element will be turned on. Additionally, our system has to check the temperature conditions every 20 s, although most other intervals may be set as well. The schematic circuit of this system is shown in Fig.2.4:

This system includes an LM35DZ temperature sensor which produces an analog voltage proportional to a temperature of an environment with a factor of 10mV/°C. For example, if the LM35's output voltage is equal to 0.29 V, the temperature of the environment is 29°C.
The sensor's output signal passes to the non-inverting input (pin 3) of the voltage comparator LM2903 (A2). The inverting input (pin 2 of A2) is tied to the reference voltage that comes from the wiper of the potentiometer R2 attached, in turn, to the +5V power supply. This reference voltage defines the threshold and may be selected randomly by a user. If the sensor's output voltage on the non-inverting input exceeds the reference voltage on the inverting input the output (pin 1 of A2) of the comparator goes high. Otherwise, if the sensor's output becomes less than the reference voltage the comparator's output goes low. The program checks the comparator's output via bit 3 of the status port (address 0x379).

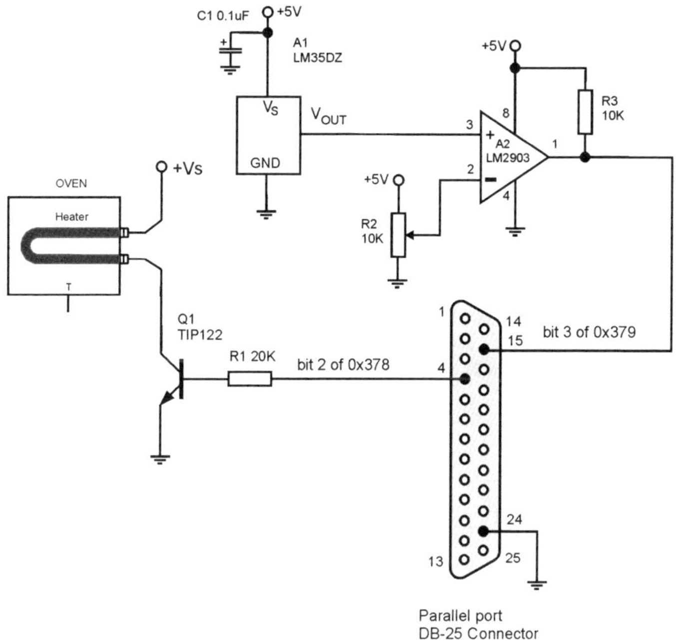

Fig.2.4

After bit 3 has been set the program code turns off the heater element by clearing bit 2 of the data port (address 0x378). In this case a low voltage level on pin 4 of the DB-25 connector cuts off the Darlington transistor TIP122 (Q1) which, in turn, stops powering the heater element. Bit 3 of the status port is set to "0" when the temperature of the oven drops below the threshold. In this case the program turns on the heater element by setting bit 2 of the data port.

The source code of the control program is shown below (Listing 2.4):

Listing 2.4

```
#include <stdio.h>
#include <sys/io.h>
#include <unistd.h>

int main(void)
{
 int dataPort = 0x378;
 int statusPort = 0x379;
 int Bit3;

 iopl(3);
 while(1)
   {
     Bit3 = inb(statusPort) & 0x8
     if (Bit3 != 0)
      outb(0x0, dataPort);
     else
      outb(0x4, dataPort);
     sleep(20);
   }
 return 0;
}
```

Let's discuss the above source code. The control algorithm is implemented inside the endless `while()` loop. Each iteration begins by reading the status port and saving the byte read in the `Bit3` variable. Then the program code checks the value of `Bit3`. A non-zero value says that the temperature has exceeded the threshold; therefore the heater is switched off by executing the statement

```
outb(0x0, dataPort);
```

A zero value of `Bit3` indicates lowering of the temperature, so the heater will be switched on:

```
outb(0x4, dataPort);
```

The last statement of each iteration is the `sleep(20)` function which performs a 20 second delay before launching the next iteration. You can set your own delay value instead of 20 if needed.
This project can be easily customized. Instead of the LM2903 chip you can use another single-supply comparator with a TTL output. If you want to reach a higher precision, you should consider a

comparator with a low offset voltage – this minimizes the error introduced by the comparator when configuring a threshold voltage.
It is worth using a high precision potentiometer R2 – which also affects the accuracy of this circuit. Actually, the best result would be achieved by applying a specialized single-chip voltage reference instead of R2. When designing this project you should also evaluate your real load driven by Q1 and select an appropriate power transistor. Additionally, common precautions must be taken into account when wiring devices to the parallel port. Lastly, if you are going to use some expansion board with the parallel interface, remember to replace addresses 0x378-0x379 in the source code with appropriate values.

Expanding input lines of the parallel port

Often you need more inputs to capture signals from external circuits than just the parallel port can provide. The simplest way to expand the number of inputs is to use a multiplexer which can direct multiple input signals to a single output. The following embedded project demonstrates how to do that. The hardware part of the project is presented in the following schematic circuit (Fig.2.5):

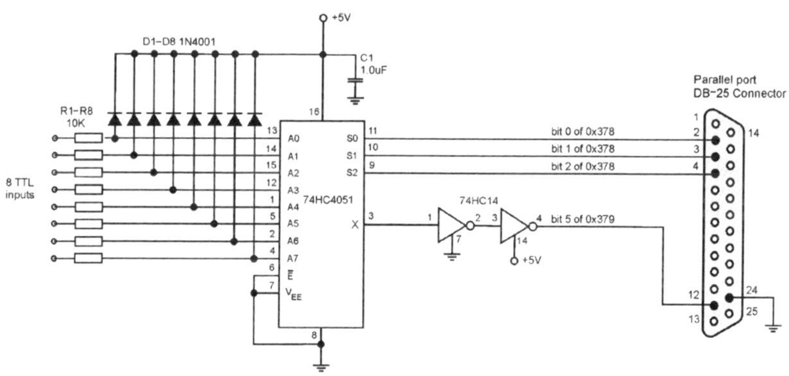

Fig.2.5

This circuit allows using up to 8 additional inputs with the parallel port. As you can see the "heart" of this circuit is a multiplexer type 74HC4051. This circuit provides reading of the A0-A7 digital inputs one at a time. The TTL-compatible signal of the selected input goes to the single X output of the 74HC4051. The X output, in turn, is attached to an input of the parallel port. In our case X goes to pin 12 of the DB-25 connector. This given pin represents bit 5 of the status port (register). Thus, the control program may know the state of the X output by reading bit 5 of the status port. As usual the control program uses an address 0x378 for the data port and address 0x379 for the status port (register). The program selects an input channel for reading by setting the 3-bit digital code on the S0–S2 select inputs of the 74HC4051. For example, setting a code 0x0 on S0–S2 allows reading of A0; code 0x1 allows reading of A1, etc. The S0–S2 inputs are controlled through bits 0–2 of the data port of the parallel port. Some words about the design. The resistor–diode pairs (RxDx, x = 1…8) protect the multiplexer's inputs from higher voltages; the 74HC14 buffer provides a valid TTL signal for the input of the status port. Some words about wiring. The length of wires connecting this circuit to the parallel port must not exceed 30 cm, otherwise you will need additional buffers to provide clean signals. Also it is necessary to provide bypass capacitance on the power supply. An adequate bypass is a 1μF capacitor connected as close as possible to the power pin of the multiplexer. As an alternative to the 74HC4051, you can use any 4051 multiplexer to be operating with TTL signals. Let's talk about software. The control application has the following source code (Listing 2.5):

Listing 2.5

```
#include <stdio.h>
#include <unistd.h>
#include <sys/io.h>
#include <stdlib.h>

int main(void)
{
 int channel;
 int dataPort = 0x378;
 int statusPort = 0x379;
 int result = 0;
 int tmp = 0;

 iopl(3);
 while (1)
 {
  for (channel = 7; channel >= 0; channel--)
  {
   outb(channel, dataPort);
   tmp = inb(statusPort) & 0x20; // selection 5th bit of the
                                 // statusPort

   tmp = tmp >> 5;
   tmp = tmp << channel;
    result = result | tmp;
  }
```

```
  printf("Input states A7---A0: 0x%X\n", result);
  sleep(2);
  result = 0x0;
 }
 return 0;
}
```

In this source code we use several variables. The `channel` variable contains the actual number of the channel that will be read. The program code sequentially decrements this variable from 7 to 0 when reading data in the `for()` loop. The `dataPort` and `statusPort` variables hold the address of the data port and status port respectively. The final value is saved in the `result` variable. The `tmp` variable holds the value of bit 5 of the status port for each channel read.

In each iteration inside the `for()` loop the program reads an appropriate channel, saves the value of bit 5 in the `tmp` variable and introduces the appropriate bit into the `result` variable. The state of all 8 inputs is output by calling the `printf()` library function. The time interval between iterations is selected to be equal to 2 s (the `sleep(2)` function). Naturally, it is up to you to set an appropriate pause for your own control application.

You can see the result of executing the program in Fig.2.6:

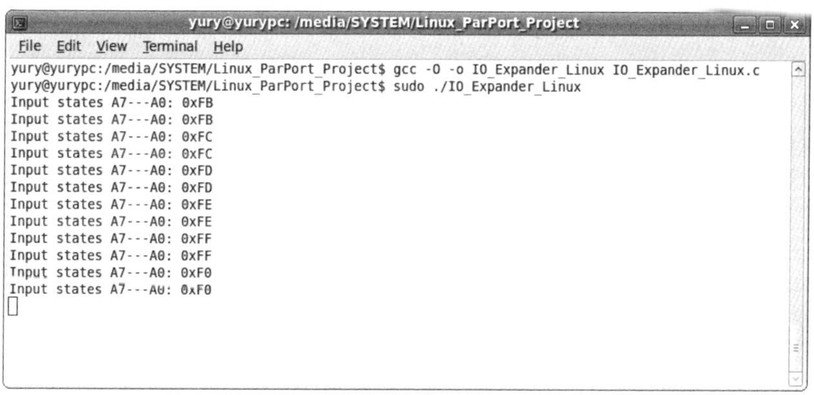

Fig.2.6

This embedded project can be easily improved by expanding the inputs up to 16. This requires using the 74HC4067 multiplexer instead of the 4051 and implementing small modifications to the schematic circuit. Also you need to slightly modify the source code for the 4051 device so that the program can also work with the 4067.

So far we discussed systems which operate mainly with digital signals. The following embedded project concerns measuring analog signals. Despite the "digital world" of electronics that surrounds us the real world of course actually operates with analog signals. These may mainly originate from a physical environment or various electronic circuits. Temperature, humidity, pressure, light and other physical quantities are continuous (analog) by nature, so all they need analog sensors and analog conditional circuits before they pass to microcontrollers or microprocessors. Even digital sensors are not "purely digital" because they use analog circuits to process the analog signals inside their modules.

Since a PC does not "understand" analog signals we need to convert them into a digital form. This process is called a "digitization". To convert (digitize) an analog signal into its binary representation, we need a special device usually called an "analog-to-digital converter" (ADC). An analog-to-digital converter captures an analog signal and converts it into a digital bit stream. This bit stream, in turn, may be "pulled out" of the ADC by the software running on a PC via some interface, such as SPI, I^2C, etc. Finally, a digital image of the analog signal may be used in the PC by high-level software for complex processing and interpretation of results.

Measuring analog signals

This project illustrates how to measure an analog signal represented as a voltage level which could be output from a typical sensor. As usual, this is a classic embedded project which includes both hardware and software. The hardware is very simple and includes a 12-bit successive approximation (SAR) analog-to-digital converter, type MCP3201, wired to the PC parallel port via a SPI-compatible interface (Fig.2.7):

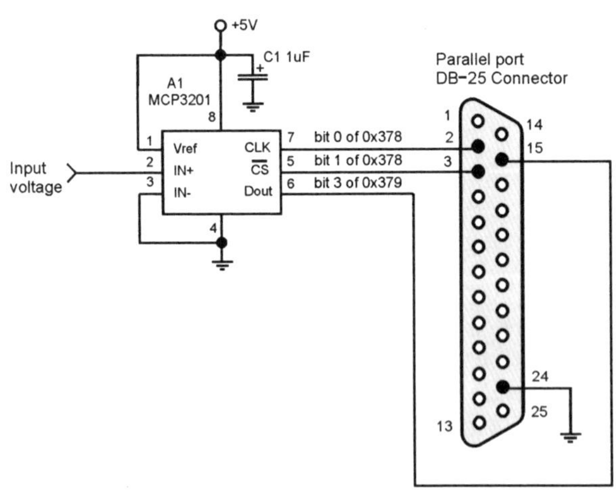

Fig.2.7

The SPI interface uses three signal lines, CS, CLK and Dout. The analog-to-digital conversion and data transfer are controlled via those lines by the program running on the PC. The analog input signal in the range of 0 to +5V is placed on input pin 2 of the MCP3201 chip. The CS control line frames the conversion process and must be low while the conversion is in progress. The program controls this line via bit 1 of the data port (pin 3 of the DB–25 connector). The output bit stream appears on the Dout line connected to bit 3 (pin 15 of the DB–25 connector) of the status port (0x379).
The software reads each bit of this bit stream (MSB first) and saves it into an integer variable. After all data bits have been captured the application takes the lowest 12 bits which represent the original analog signal. The CLK line provides the clock pulses thus synchronizing the data transfer. These pulses appear on pin 2 (bit 0 of the data port). Each separate bit is set on Dout on the falling edge of CLK and should be latched by the control application on a rising edge of the clock pulse. To complete the conversion process the program must raise the CS line.
What was just said is illustrated by the timing diagram shown in Fig.2.8:

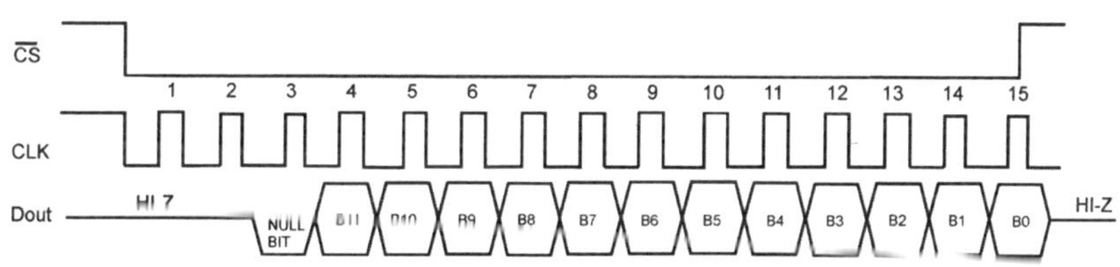

Fig.2.8

Note, that the MSB bit appears first on the Dout line after the 3rd CLK pulse has gone low. As you can see, a data transfer cycle requires 15 iterations.
The circuit should be powered from a +5V noiseless power supply. The bypass capacitor of 10 uF should be placed as close as possible to the power pin of the MCP3201 chip. The MCP3201 device may be replaced using a similar successive approximation ADC with an SPI-compatible interface. For instance, you may use a LTC1286 or a LTC1297 device (Linear Tech.), etc. In this case you would need to reconsider the circuit connections and modify the source code according to the timing diagram of the device selected.
Let's discuss the software. The source code of our control application is presented below (Listing 2.6):

Listing 2.6

```
#include <stdio.h>
#include <unistd.h>
#include <stdlib.h>
#include <sys/io.h>

#define dataPort   0x378
#define statusPort 0x379

int main(void)
{
 int i1;
 int binData = 0;
 int mask = 0x4000;
 int dataStatusPort = 0;
 double total = 0;

 iopl(3);                // CS   is a Bit 1 of dataPort,
                         // CLK  is a Bit 0 of dataPort,
                         // Dout is a Bit 5 of statusPort
 outb(0x2, dataPort); // CS goes high
 outb(0x3, dataPort); // CS sticks high, CLK goes high
 outb(0x1, dataPort); // CS goes low thus allowing the conversion
 for (i1 = 0; i1 < 15; i1++)
 {
  outb(0x0, dataPort);                    // CLK goes low
  dataStatusPort = inb(statusPort) & 0x8;
  if (dataStatusPort != 0)
    binData = binData | mask;
  outb(0x1, dataPort);                    // CLK goes high
  mask = mask >> 1;
 }
 outb(0x3, dataPort);
 binData = binData & 0xFFF;
 total = 5.08 / 4096 * binData;
 printf("MCP3201 gives following result: %5.3f V\n", total);
 return 0;
}
```

First of all, let's examine the variables used. The `binData` variable collects the binary code bit by bit. After the last iteration has been completed this value will hold a binary code representing the actual analog signal. The `total` variable of a double type holds the final value of the analog voltage; this value will be output onto the screen by the `printf()` function. Look at the following statement:

```
total = 5.08 / 4096 * binData;
```

Here the constant of 5.08 is equal to the reference voltage (in volts) applied to pin 1 of the MCP3201 converter. This reference voltage comes from my specific power source; maybe you will use another value of reference voltage for your own circuit. The value of the reference voltage divided by 4096 (2^{12}) gives the size (in volts) of the LSB of the converter. In our case this value will be 5.08 / 4096 = 0.00124V (1.224mV). The program multiplies the binary code held in `binData` by the LSB and saves the final value of the input voltage in the `total` variable.

The `dataStatusPort` variable holds a bit captured from the bit stream outpu by the ADC. This value is used when the program writes a particular data bit into the appropriate position in the `binData` variable.

The control application implements the timing diagram of the conversion inside the `for()` loop. Since the full conversion process requires 15 iterations, the program code sequentially reads 15 bits. Only the first 12 lowest order bits are of interest to us, therefore we throw away a higher 3 bits by executing the statement

```
binData = binData & 0xFFF;
```

After typing the source code we can save it in the ADC.c file and compile the source by using the command:

```
$ gcc -o ADC ADC.c
```

To launch the application we would enter:

```
$ sudo ./ADC
```

The projects discussed may be used as templates for designing more complex systems. Now it is time to discuss analog conditioning circuits.

Processing analog signals: common technique

In the previous project we applied the analog–to–digital converter to process an analog signal. Such a simple digitization system exhibits good results in many home embedded projects with a low-level of noise. If you apply a low-noise analog signal from your sensor and your power source doesn't generate much noise, it is reasonable to connect an analog-to-digital converter's input directly to the signal source. Many practical applications where such conditions are met will exhibit good performance and accuracy.

Unfortunately, in real life any useful analog signal has some noise wedged in,.We need to get rid of this unwanted component by applying some additional circuitry. From the simplest point of view noise may be regarded as a combination of various random signals originated from various external and internal sources. In general, designing a noiseless circuit is a very complicated task, so we will not discuss this topic here. We will only consider how to attenuate or except noise components from analog conditioning circuits in practice. The simplest and effective way to get rid of noise from a signal of interest is based on using analog low-pass filters (LPF).

Often we deal with analog signals which fall into the full-scale range (FSR) of the analog-to-digital converter. Sometimes signals are too small so they must be amplified before entering the converter's input. Putting all this together, we can figure the general scheme of an analog conditioning chain (Fig.2.9) – this may be useful in understanding how analog signals may be processed in digital systems. As you can see, at the first stage the sensor's signal may be increased by an **Amplifier** block. Using this stage achieves the only goal of increasing the magnitude of an original signal so that it would match the FSR of the following analog-to-digital converter.

The **Amplifier** block is usually built on either an operational ("op amp") or instrumentation amplifier, depending on the type of the output signal that the particular sensor provides. You can find a lot of information about op amps and their use on the Internet. In this book guide we don't use the **Amplifier** stage in our projects, although you may need to do that in your own projects.

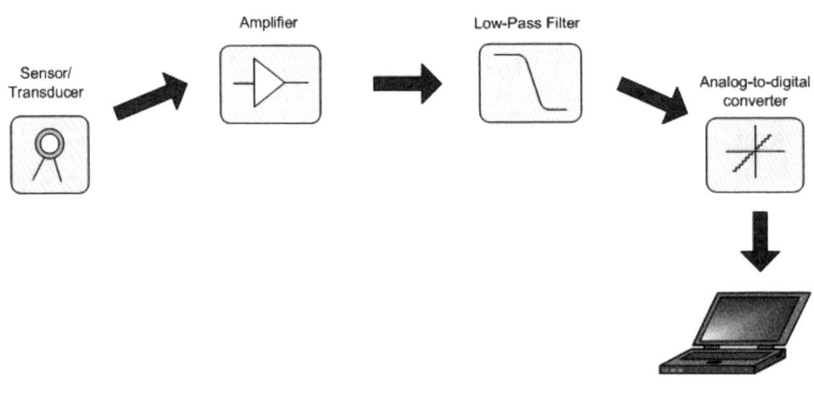

Fig.2.9

An amplified analog signal is fed into an analog low–pass filter (**Low-Pass Filter** in Fig.2.9). The analog low–pass filter (LPF) removes or at leastattenuates out-of-band noise riding on the useful signal. Since all analog signals have some noise, the filter stage is almost always required. You can, however, accept some reasonable degree of noise in your particular system when it doesn't affect the results you expect. In these cases you can exclude the low-pass filter stage from the signal chain. The low-pass filter always precedes the analog–to–digital converter (**Analog–to–digital converter** in Fig.2.9), because the noise must be suppressed before it appears at the converter's input. If the signal of interest remains coupled with a noise signal, an analog-to-digital converter provides a "dirty" binary code which cannot be completely cleaned, even with digital filters. After digitization the binary representation of an analog signal may be processed in a "digital world" by a microcontroller or processor of a PC.

For analog signals with relatively high voltage levels you can apply a successive approximation register (SAR) analog-to-digital converter having a relatively wide FSR. If the sensor yields a low–level output signal, a developer either may amplify the signal or utilize some kind of a Σ–Δ (sigma–delta) analog–to–digital converter that can provide 16 to 24–bit resolution. Using Σ–Δ ADC gives an additional financial benefit, because some electronic components may be excluded from your design. However, you should remember that Σ–Δ converters are still slow compared to SAR devices, so Σ–Δ often function in low–speed measuring systems. The SAR converter, on the other hand, provides a lower resolution (typically 12 bit) compared to Σ–Δ, but has a higher conversion speed. All the projects described in this book are based on using SAR analog–to–digital converters. Both types of converters (SAR and Σ–Δ) provide either SPI or I^2C interface for the interaction with micro-controllers or processors.

Let's see in brief how an analog low–pass filter functions. Recall, that a LPF weeds out higher frequency noise coupled with a useful signal. The simplest low–pass filter shown in Fig.2.10 is a simple RC pair (resistor and capacitor):

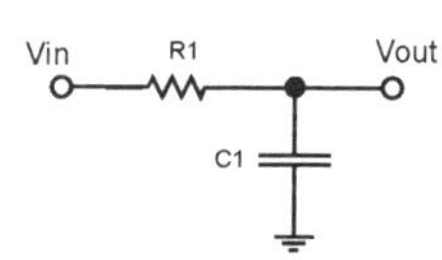

Fig.2.10

Such a filter will attenuate noise frequencies above $f = 1/(2\pi \times R1 \times C1)$. The frequency determined by this expression is frequently called a "cut–off" frequency. The point where the frequency is equal to
$1/(2\pi \times R1 \times C1)$ is called the "pole"; the pole indicates the point where signals with a frequencies higher than $1/(2\pi \times R1 \times C1)$ start to drop. In many cases the low–pass filter is designed so that it could attenuate the 50 or 60 Hz noise coming from the power supply. The cut–off frequency of the filter may be chosen very low (some Hz or less).
The higher efficiency may be reached by using so–called "multi-pole" filters based on an active filter topology. Such filters are known as "active filters"; they usually include at least one operational amplifier and some resistors and capacitors. One of the popular low-pass active filter topologies known as a "Sallen–Key" filter is given in Fig.2.11:

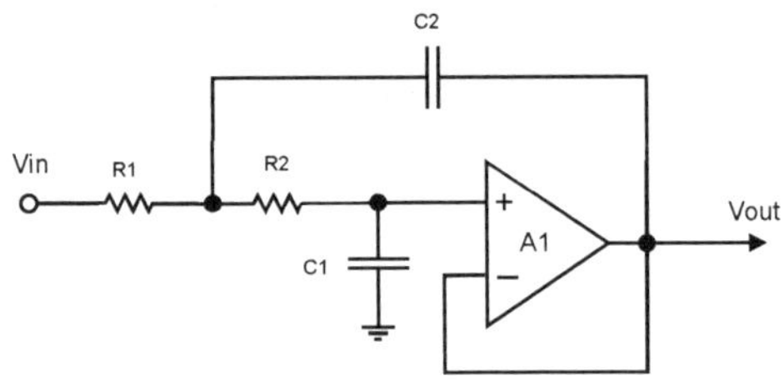

Fig.2.11

Compared to the simplest RC–filters, active filters have better characteristics. Additionally, active filters allow buffering of a high–impedance signal source from an analog–to–digital converter's relatively low impedance input. Also active filters may have a gain above 1V/V that allows amplifying small analog signals.

The low-pass active filter shown in Fig.2.11 is, in fact, a voltage follower with a gain of 1V/V in the pass-band. Since there are two capacitors the filter has two poles (2nd order filter). Higher order filters can improve the frequency response, although their schematic circuit is more complicated compared with the simplest filter topology.
In practice, the 2nd order low–pass filter will be a good choice for most applications. The analog signal chain with the Σ–Δ analog–to–digital converter may include the simplest low-pass filter based on a single RC–pair (see Fig.2.10) instead of an active LPF.

In simple circuits with a low noise it is possible to use the non-inverting low-pass filter, being implemented as a voltage follower with a single RC pair (Fig.2.12):

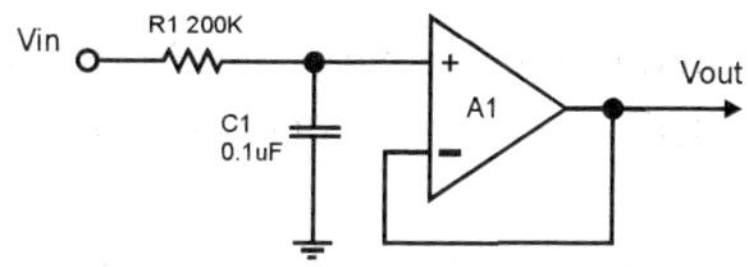

Fig.2.12

You can also use a simple inverting low-pass active filter shown in Fig.2.13:

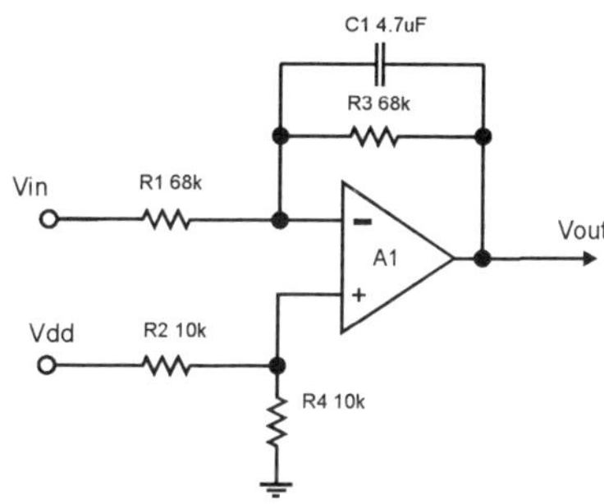

Fig.2.13

As absolutely "clean" analog signals do not exist, some sort of the noise will exist.. It is up to you whether to ignore noise or not, although using low-pass filters can prevent many problems when designing analog conditioning circuits.

Unlike analog sensors, their digital counterparts provide a "pure" binary code on their outputs. This code may be additionally processed and analyzed by a micro-controller or processor. This significantly simplifies the design of embedded systems because conditioning circuits may be simplified or even excluded.

A digital sensor usually provides amplifying, filtering and a digitization of signals inside the chip. Digital sensors have their own built-in digitizing system (SAR or, more frequently, sigma-delta converter) and may contain some additional circuitry such as filters and amplifiers. The developer usually configures operating modes of the digital sensor by writing some values to inner registers before starting a measurement. Afterwards data being measured may be read into the microcontroller or processor according to a device's timing diagram.

Most digital sensors have an interface compatible with the SPI or I^2C standard protocols. Though using digital sensors simplifies the design, there are, however, some drawbacks. It is still difficult, for example, to gain a high precision for these sensors when a physical quantity changes over a wide range. Anyway, the choice of sensor is up to the developer.

Let's keep on discussing projects. The following project describes a simple cooling system where the LM35DZ temperature sensor together is involved.

Simple cooling system

The hardware part of our cooling system is shown in Fig.2.14 (a):

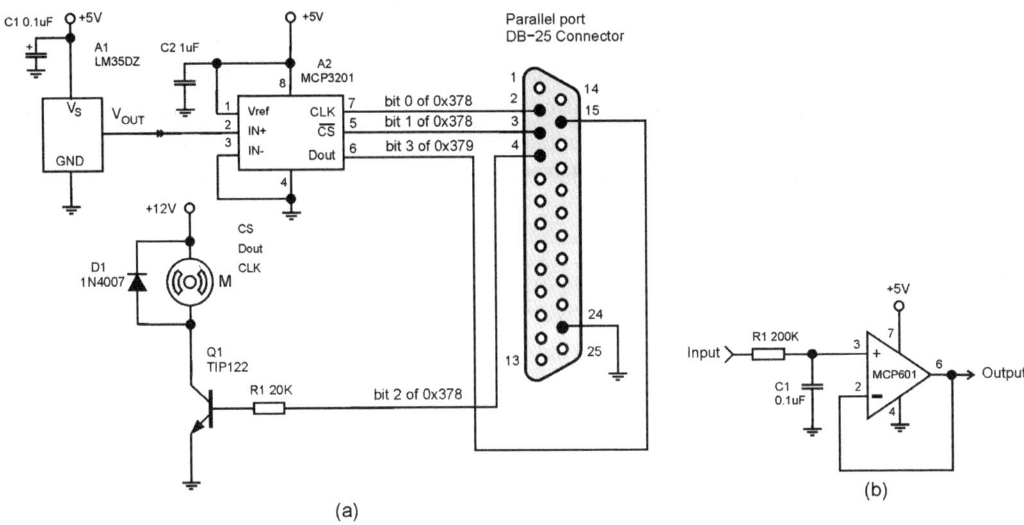

Fig.2.14

As you can see, the temperature in the room is taken from the LM35DZ temperature sensor which provides an analog voltage proportional to the temperature of the environment with a factor of 10mV/°C. This gives, for instance, the LM35DZ output voltage of 0.21V at an environment temperature equal to 21°C.

As you can see, the output signal of the LM35DZ falls into the FSR of the 12-bit analog-to-digital converter MCP3201, so we can directly pass the signal to the ADC's input (pin 2 of MCP3201). After the conversion process has been completed the binary representation of the analog signal is transferred to a PC parallel port via a SPI-compatible interface. Also we assume that the signal coming from the LM35DZ is free of noise, therefore we don't use any low-pass filter before the analog-to-digital converter's input.

This system operates in the following way. When the temperature of the environment where the sensor has been placed exceeds some threshold, the program turns on the motor which starts to operate the cooler. When the temperature drops below the threshold, the motor will be off, so the cooler

stops running. For this project the threshold is set at 0.23V, corresponding to 23°C.
The motor is controlled via bit 2 (pin 4) of the data port. The motor will be turned on by raising a voltage level on pin 4 of the DB–25 connector. The high level on pin 4 turns on the Darlington transistor TIP122 (Q1) which, in turn, switches on the motor. The motor will be off when the voltage on pin 4 goes low. In this case the low-level voltage cuts off the Q1 transistor which, in turn, turns off the motor.
If the LM35DZ's signal is contaminated by noise, you should introduce some low-pass filter to remove unwanted signal components. The filter may be either a simple RC-pair or a more effective active LPF where the circuit is shown in Fig.2.14 (b). In this case the "Input" of the filter is connected to the output of the sensor, while the "Output" goes directly to pin 2 of a MCP3201. As you can see, this active low-pass filter is very simple and consists of the MCP601 op amp and the R1C1 pair.
Let's go to the software. Our control application, in general, could be responsible for managing our cooling system round-the-clock without intervention of a user. The best variant to accomplish this may be realized by a special type of application that is known as a "daemon". A UNIX/Linux daemon (service) is a background process that is running autonomously, with little or no user intervention.

Many important processes in Linux are handled by daemons. Although writing a daemon may seem difficult and slightly mysterious, in real life even if you are not an experienced programmer or developer you can still write a simple daemon to handle the particular embedded system. The following C source code (Listing 2.7) represents the daemon which controls our cooling system.

Listing 2.7

```
#include <stdio.h>
#include <unistd.h>
#include <stdlib.h>
#include <sys/io.h>
#include <strings.h>

#define dataPort   0x378 // data port
#define statusPort 0x379 // status port

int main(void)
{
 pid_t PID;
 FILE *fo;
 char buf[4096];
 int bRead;
 int i1;
 int binData;
```

```
int mask;
int dataStatusPort = 0;
int OnOff = 0;
double total = 0;

PID = fork();
if (PID)
  exit(0);
umask(0);
setsid();
chdir("/");
  printf("Cooling   system   started   with   a   PID   =   %d\n",
(int)getpid());
close(0);

close(1);
close(2);
unlink("/home/yury/templog");
iopl(3);                 // CS   is a Bit 1 of dataPort,
                         // CLK  is a Bit 0 of dataPort,
                         // Dout is a Bit 5 of statusPort
while (1)
{
 mask = 0x4000;
 binData = 0x0;
 total = 0;
 outb(0x2 | OnOff, dataPort); // CS goes high
 outb(0x3 | OnOff, dataPort); // CS sticks high, CLK goes high
 outb(0x1 | OnOff, dataPort); // CS goes low thus
                              // allowing the conversion
 for (i1 = 0; i1 < 15; i1++)
  {
   outb(0x0 | OnOff, dataPort);            // CLK goes low
   dataStatusPort = inb(statusPort) & 0x8;
   if (dataStatusPort != 0)
     binData = binData | mask;
   outb(0x1 | OnOff, dataPort);            // CLK goes high
   mask = mask >> 1;
   }
  outb(0x3 | OnOff, dataPort);
  binData = binData & 0xFFF;
```

```
    total = 5.08 / 4096 * binData;

    // Writing result to the file

    bzero(buf, sizeof(buf));
    fo = fopen("/home/yury/templog", "a+");
    bRead = sprintf(buf, "%s", "LM35DZ output: ");
    bRead += sprintf(buf + bRead, "%5.3f %s\n", total, "V");
    fwrite(buf, sizeof(char), bRead, fo);
    fclose(fo);
    if (total > 0.23)
    {
     OnOff = 0x4; // bit 2 of data port goes high to turn on the motor
    }
    else
    {
     OnOff = 0x0;
    }
   outb (0x2 | OnOff, dataPort);
   sleep(20);
  }
 return 0;
}
```

Creating a daemon in Linux uses a specific set of rules in a given order. Daemons should never have direct communication with a user through a terminal. In general, a daemon should not communicate directly with a user at all. All communication should pass through some sort of interface. Our daemon will simply write a value of temperature measured into a text file named as `templog` that is created in a `/home/yury` directory (you can use other name and directory). The daemons will repeat every 20 s measuring the temperature and add information obtained into the `templog` file. The content of this file then may be analyzed later by the user, if necessary.

When a daemon starts up, it has to do some preparations to get itself ready for its real job. First of all, the daemon should fork off the parent process. This is done by executing the following statements:

```
PID = fork();
if (PID)
  exit(0);
```

In this section of code the `fork()` function returns the process id (PID) of the child process. If PID returned is not equal to zero, we must exit the parent process by calling the `exit()` function. Equality of PID to -1 requires that the program should terminate the daemon right here. For clarity we don't analyze the case when PID is equal -1, so it is up to you to add some checking in the source code. The next step we need is to change file mode mask by using an umask() API call:

```
umask(0);
```

After this statement has been executed we can write to/read from any files created by the daemon. By setting umask to 0, we will have full access to the files created by the daemon. In our case, specifically we will be able to read the /home/yury/templog file.
The next step allows getting a unique SID for the child process; this grants the process will operate properly. Otherwise, the child process becomes an "orphan" in the system. To accomplish this the program should call the setsid() function. Then we need to change the current working directory to a safe place. The current working directory should be changed to some place that is guaranteed to always be there. The only directory that meets this condition is the root (/). Changing the working directory is accomplished by invoking a chdir() call:

```
chdir("/");
```

One of the last steps when setting up a daemon is closing out the standard file descriptors denoted as STDIN (equal to 0), STDOUT (equal to 1) and STDERR (equal to 2). Since a daemon cannot use the terminal, these file descriptors are redundant and a potential security hazard. The following section of code does this:

```
close(0);
close(1);
close(2);
```

By now we have almost configured the daemon's code. At this point, we have basically told the Linux operating system that this program code will run as a daemon, so we can introduce an actual daemon's code.
Each time we start our daemon we want the data to be written into a newly created text file (I called it /home/yury/templog). Before doing this the program code should try to delete the previously created templog file (if a given file still exists) by executing the following statement:

```
unlink("/home/yury/templog");
```

As we want to operate with hardware registers, we need to aquire the appropriate permissions to do that by executing the iopl() function:

```
iopl(3);
```

Then we go into the endless while() loop where the main daemon's job is performed. We must accomplish two tasks inside the `while()` loop: digitize the LM35DZ analog signal and get the result written into the `templog` file. The digitization (conversion) is implemented inside the `for()` loop (we discussed this earlier). After the conversion has ended the program code forms the text representation of the result and writes the text string so created into the `templog` file. The following section of code does that:

```
bzero(buf, sizeof(buf));
fo = fopen("/home/yury/templog", "a+");
bRead = sprintf(buf, "%s", "LM35DZ output: ");
bRead += sprintf(buf + bRead, "%5.3f %s\n", total, "V");
fwrite(buf, sizeof(char), bRead, fo);
fclose(fo);
```

Here we create a new file with an `fopen()` library function and get the pointer to the `FILE` object saved in the `fo` variable; this variable will be used for the following file operations. The text string is formed using the `sprintf()` library function which utilizes an array of bytes `buf` to hold there the resulting string. The `fwrite()` library function writes the string just created into the `fo` variable. Finally, the `fo` object is closed by calling the `fclose()` function.
The following statements implement a switching on/off of the cooler after the threshold value (that is set to 0.23) has been checked:

```
if (total > 0.23)
 {
  OnOff = 0x4;
 }
else
 {
  OnOff = 0x0;
 }
outb (0x2 | OnOff, dataPort);
```

Each iteration is ended with the `sleep(20)` function which executes the delay. That's about all there is to say about this control application. Save our source code in the file LM35DZ_Cooler.c and compile it by entering the following command:

```
$ gcc -o LM35DZ_Cooler LM35DZ_Cooler.c
```

If the compilation was successful we can start our daemon:

```
$ sudo ./LM35DZ_Cooler
Cooling system started with a PID = 9733
```

It's worth checking our process with the `ps` command:

```
$ ps -ef | grep LM35DZ_Cooler
root       9733     1  0 19:37 ?         00:00:00 ./LM35DZ_Cooler
yury       9736  9516  0 19:37 pts/1     00:00:00 grep LM35DZ_Cooler
```

After some time has passed we can test the results of measurements written to the text file by using, for example, a `cat` command:

```
$ cat /home/yury/templog
LM35DZ output: 0.196 V
LM35DZ output: 0.201 V
LM35DZ output: 0.238 V
LM35DZ output: 0.238 V
LM35DZ output: 0.274 V
LM35DZ output: 0.275 V
LM35DZ output: 0.275 V
LM35DZ output: 0.255 V
LM35DZ output: 0.241 V
LM35DZ output: 0.227 V
```

As the `templog` file is usual text, its content may be viewed using some other appropriate Linux command or editor (vi, ed, gedit, etc.) as well.
To stop our daemon we need to enter the `kill` command:

```
$ sudo kill -9 9733
```

You could also use "sudo pkill `LM35DZ_Cooler`"

This embedded project can be adapted to create any simple system where some physical conditions (temperature, humidity, pressure, light, etc.) should be held unchanged.
Next projects describe systems where digitally controlled electronic circuits are involved. Let's see how to build and use a wide-band signal generator which can function in a range of 1 KHz–68 MHz.

Wide-band digitally controlled oscillator

This project describes a simple wide-band signal generator that can issue a square wave in the range of about 1 KHz through 68 MHz. This generator is based on a popular low-cost chip LTC6903 from Linear Technology. Some words about how the oscillator's chip functions.

The LTC6903 device is a low-power, self contained digital frequency source providing a precision frequency from 1 KHz to 68 MHz, set-able through a serial port. The LTC6903 requires no external components other than a power supply bypass capacitor, and it operates over a single wide supply range of 2.7V to 5.5V. In our project the supply voltage will be 5.0V.
As the datasheet says, the LTC6903 features a proprietary feedback loop that linearizes the relationship between the digital control setting and the output frequency, resulting in a very simple frequency setting equation:

$$f = 2^{OCT} \times 2078(Hz) / (2 - DAC/1024) \quad (1),$$

where 1 KHz < f < 68 MHz, OCT is a 4-bit integer value (0-15) represented by the serial port register bits OCT [3:0] and DAC is a 10-bit integer value (from 0 through 1023) represented by the serial port register bits DAC[9:0].
The output frequency can be set programmatically via a 3-wire SPI-compatible interface as shown in Fig.2.15:

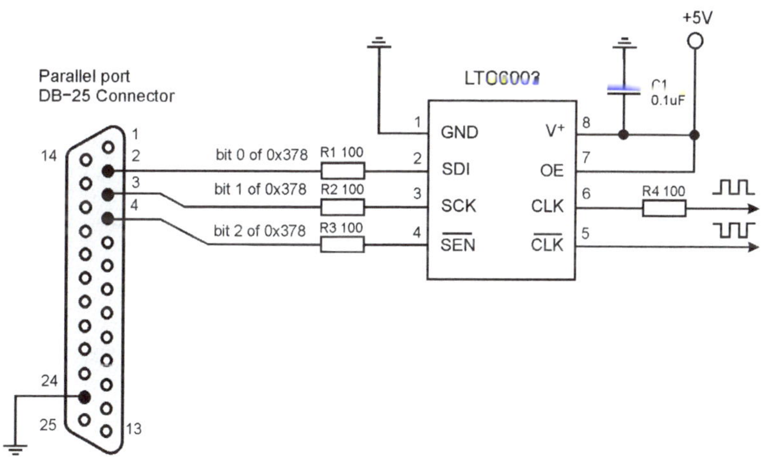

Fig.2.15

The SPI signal lines are controlled via bit 0 through bit 2 of the data port (0x378). The output frequency appears on the CLK output (pin 6) or on pin 5 (the reversed signal). The OE signal line attached to the source voltage allows both CLK outputs. In order to both lessen the noise and increase the accuracy a bypass capacitor of about 0.1uF should be tied to the power pins of the LTC6903 chip as close as possible. It is also desirable to cut off any excess of the capacitor's leads as much as possible to minimize their series inductance
Let's see how to configure the LTC6903's output frequency.

The appropriate output frequency of LTC6903 can be set by using the formula (1). Configuring the frequency involves the following steps:

1. Selection of the appropriate value of OCT from the Table 2 shown below:

Table 2. Output Frequency Range vs. OCT Setting (Frequency Resolution 0.001 * f)

f ≥	f <	OCT
34.05 MHz	68.03 MHz	15
17.02 MHz	34.01 MHz	14
8.511 MHz	17.01 MHz	13
4.256 MHz	8.503 MHz	12
2.128 MHz	4.252 MHz	11
1.064 MHz	2.126 MHz	10
532 KHz	1063 KHz	9
266 KHz	531.4 KHz	8
133 KHz	265.7 KHz	7
66.5 KHz	132.9 KHz	6
33.25 KHz	66.43 KHz	5
16.62 KHz	33.22 KHz	4
8.312 KHz	16.61 KHz	3
4.156 KHz	8.304 KHz	2
2.078 KHz	4.152 KHz	1
1.039 KHz	2.076 KHz	0

2. Choosing the variable DAC using the following formula, and rounding DAC to nearest nearest integer:

$$DAC = 2048 - 2078(Hz) \times 2^{(10+OCT)} / f \qquad (2)$$

3. After OCT and DAC are selected you can form the 16-bit binary code that the program writes to the serial register of the LTC6903 chip (Fig.2.16):

D15	D14	D13	D12	D11	D10	D9	D8	D7	D6	D5	D4	D3	D2	D1	D0
OCT3	OCT2	OCT1	OCT0	DAC9	DAC8	DAC7	DAC6	DAC5	DAC4	DAC3	DAC2	DAC1	DAC0	0	0

Fig.2.16

For this particular example we assign zero to the D1–D0 fields. Once the 16-bit code is determined, we can write this code into the LTC6903 device according to the timing diagram shown in Fig.2.17:

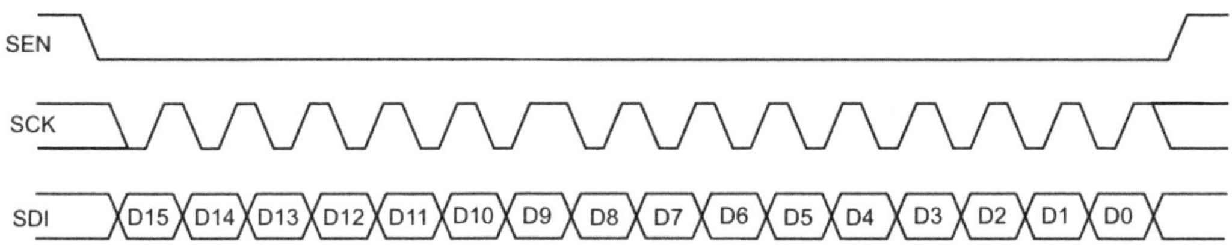

Fig.2.17

As you can see from Fig.2.17, the SEN signal frames the writing of the bit stream into the serial register of the LTC6903. When SEN goes low the data transfer is enabled. The application outputs each bit on the SDI line (MSB goes first) and writes this bit to the device on the rising edge of the SCK pulse. After the last bit (D0) has been written SEN goes high.
The timing diagram of oscillator is implemented in the control application with the following source code (Listing 2.8):

Listing 2.8

```
#include <stdio.h>
#include <unistd.h>
#include <sys/io.h>
#include <stdlib.h>

#define dataPort 0x378
#define SEN 0x4     // bit 2 of dataPort = 1
#define SDI 0x1     // bit 0 of dataPort = 1
#define SCK 0x2     // bit 1 of dataPort = 1
```

```
int main(void)
{
 int freq, tmp;
 int i1;
 int MSB;

 //freq = 0x844,  F = 1400 Hz
 //freq = 0x1774, F = 2710 Hz
 //freq = 0x1C54, F = 3380 Hz
 //freq = 0x2274, F = 4500 Hz

  freq = 0x262C; // F = 5150 Hz
  tmp = freq;
  iopl(3);
  outb(SEN, dataPort); // SEN goes high
  outb(0x0, dataPort); // SEN goes low
  for (i1 = 0; i1 < 16; i1++)
  {
   MSB = tmp & 0x8000;
   if (MSB != 0x0)
   {
    outb(SDI, dataPort);
    outb(SCK | SDI, dataPort);
   }
   else
   {
    outb(0x0, dataPort);
    outb(SCK, dataPort);

   }
  tmp = tmp << 1;
  outb(0x0, dataPort);
  }
 outb(SEN, dataPort);
 return 0;
}
```

Let's dig into this source code. As you can see, here we deal only with the data port (0x378). The binary code for configuring the frequency is saved in the `freq` variable. The timing diagram is implemented in the `for()` loop which runs 16 iterations. In our particular case we configure the device for an output frequency of 5150Hz by assigning the appropriate value to `freq`:

```
freq = 0x262C; // F = 5150 Hz
```

Other possible values for particular frequencies are shown in commented lines. The value held in `freq` is moved into the `tmp` variable. Each bit of this variable is processed in the `for()` loop. The signal SEN frames the configuring. When writing data bits into the device SEN should be held low. After the last bit has been written into the device the SEN line must be raised.
You may experiment with the oscillator's chip by setting different frequencies. More information about configuring the LTC6903 device can be found in the supplier's datasheet. The compilation of the source code saved in the LTC6903_Osc.c file may be implemented in the usual way:

```
$ gcc -o LTC6903_Osc LTC6903_Osc.c
```

To start the application use the following command:

```
$ sudo ./LTC6903_Osc
```

Digitally controlled DC voltage source

Many embedded systems require digitally controlled DC voltages for the controlling and testing of various devices such as robotics, motors, etc. Often such a DC voltage is also useful for building voltage–controlled circuits such as amplifiers and oscillators. The obvious way to create a digitally controlled DC voltage is to use some type of digital-to-analog converter (DAC). This project is dedicated to building simplest simple 8-bit digital-to-analog converter based on the MCP41010 digital potentiometer from Microchip Corp. To begin with, let's discuss how this digital potentiometer works. Some excerpts taken from the device's datasheet will help us.

According to the datasheet, the MCP41XXX/42XXX devices constitute 256 position single and dual digital potentiometers that can be used in place of standard mechanical potentiometers. Digital potentiometers have resistance values of 10K, 50K and 100K (ohms). In our project we apply a MCP41010 device which has a maximum resistance of 10K. Each potentiometer is made up of a variable resistor and an 8-bit (2^8 = 256 positions) data register that determines the wiper position. There is a nominal wiper resistance of 52 Ohm for the 10K version, 125 Ohm for the 50K and 100K versions. The resistance between the wiper and either of the resistor endpoints varies linearly according to the value stored in the data register.

The MCP41XXX series provides 256 resistor taps and accepts a power supply from 2.7 through 5.5V. In our project we will feed the MCP41010 device with +5V. This also means that the LSB will be equal to $5 / 2^8 = 0.0195$ V. If, for example, an application writes a binary code equivalent to 100 (decimal) to the device, the digital potentiometer yields $0.0195 \times 100_{10} = 1.95$ V at its output.
Let's go back to our project. The schematic circuit of the project is shown in Fig.2.18:

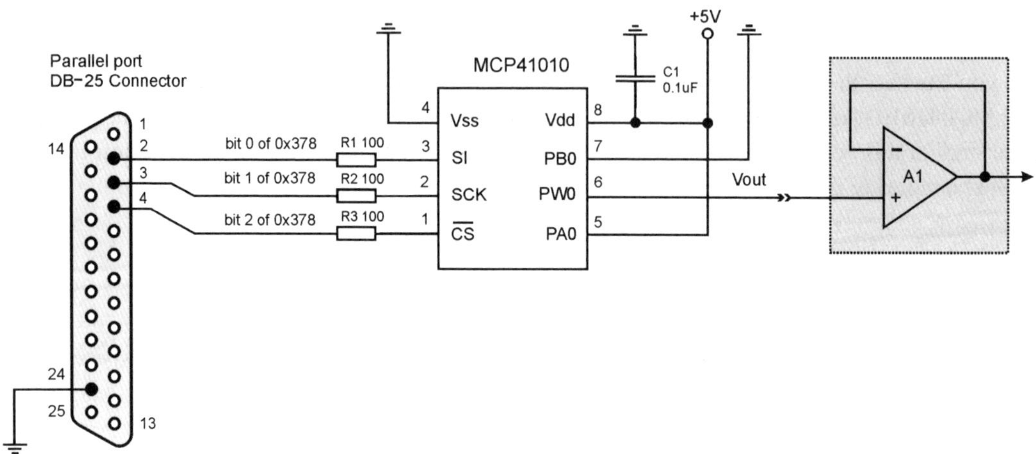

Fig.2.18

The digital potentiometer is controlled via a SPI-compatible interface from the parallel port of the PC. Three lines (SCK, SI and CS) are controlled by the application, running in Linux. The binary code that corresponds to the particular DC output voltage is shifted bit by bit from the port data register (0x378) to the digital potentiometer. The output analog voltage Vout appears on the PW0 line (pin 6 of MCP41010) that represents the "wiper" of the potentiometer.
As we apply the MCP41010 device, the total resistance between PA0 and PB0 outputs is 10K. If your circuit needs to provide a low output impedance and/ or drive a higher current load, you must use an additional buffer at the Vout output. For example, you may apply an op-amp based voltage follower (shadowed box in Fig.2.18).

In our design we use the digital potentiometer MCP41010 device in the so-called "potentiometer mode" where all three terminals of the device are tied to different nodes in the circuit. This allows the potentiometer to output a voltage proportional to the input voltage. This mode is sometimes referred to as a "voltage divider" mode. The variable voltage Vdiv is provided by adjusting the wiper (PW0) position between two endpoints PA0 and PB0 (Fig.2.19). Note that reversing the polarity of the PA0 and PB0 terminals will not affect operation.

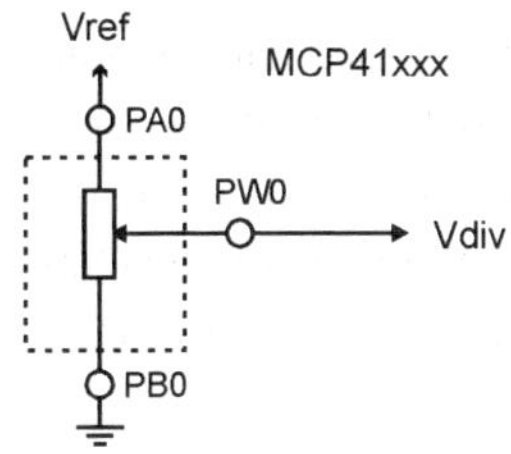

Fig.2.19

As you can see, the voltage divider consists of PA0–PW0 and PW0–PB0 resistors. In the case of the MCP41010, the full resistance between the PA0 and PB0 nodes is 10K. The resistance of each "resistor" of this voltage divider may be calculated according to the following formulas:

$$R_{WA}(Dn) = (R_{AB}) \times (256 - Dn) / 256 + R_W$$
$$R_{WB}(Dn) = (R_{AB}) \times (Dn) / 256 + R_W,$$

where R_{WA} is a resistance between the Terminal PA0 and the wiper (see Fig.2.19), R_{WB} is a resistance between the Terminal PB0 and the wiper, R_W is a wiper resistance, R_{AB} is the overall resistance for particular potentiometer (10K, in our case) and Dn is the decimal equivalent of an 8–bit binary code that determines the output voltage of the potentiometer. In our project we ignore the wiper resistance because it is very low, though for very precision circuits this value should be considered. When the device is first powered on, the data registers will be set to mid–scale (0x80).
Let's discuss how to program the MCP41010 chip. In order to provide an appropriate output voltage from the device we need to write a 16-bit value into the data register of the chip; the following timing diagram shows how to configure the digital potentiometer (Fig.2.20):

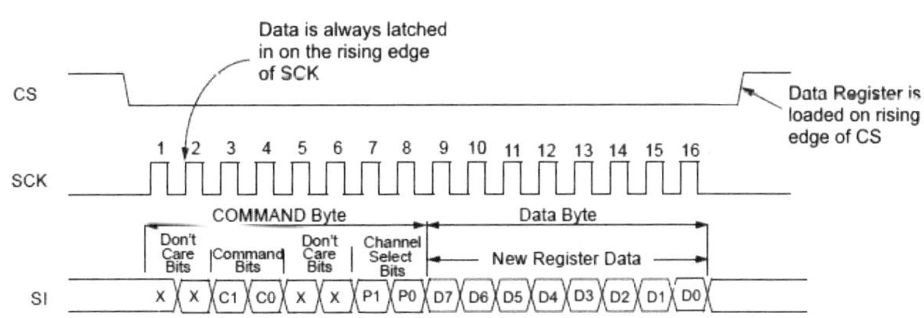

Fig.2.20

As you can see, the 16-bit code consists of command and data bytes. In our design we will set the command byte equal to 0x11, while the data byte will be the digital code (Dn) for the output voltage. For more detail you can explore the appropriate datasheet where there is a lot of information about configuring the device.

Let's discuss software. The full C source code of the control application is shown in Listing 2.9.

Listing 2.9

```
#include <stdio.h>
#include <unistd.h>
#include <stdlib.h>
#include <sys/io.h>

#define dataPort 0x378
#define CS    0x4           // bit 2 of dataPort = 1
#define SI    0x1           // bit 0 = 1
#define SCK   0x2           // bit 1 = 1

int main(void)
{
 int tmp;
 int MSB;
 int Dn;
 int VCode = 0x1100;
 int i1;

 // Dn = 32,  1/8 of the full scale  --> 0.635 V (5.08 Vref)
 // Dn = 64,  2/8 of the full scale  --> 1.27  V
 // Dn = 96,  3/8 of the full scale  --> 1.905 V
 // Dn = 128, 4/8 of the full scale  --> 2.54  V

 // Dn = 160, 5/8 of the full scale  --> 3.175 V
 // Dn = 192, 6/8 of the full scale  --> 3.81  V
 // Dn = 224, 7/8 of the full scale  --> 4.445 V

 Dn = 100; // Vout = 1.98V at Vref = 5.08V
 tmp = VCode + Dn;
 iopl(3);
 outb(CS, dataPort);         // CS goes high
 outb(0x0, dataPort);        // CS goes low to start the conversion

 for (i1 = 0; i1 < 16; i1++)
 {
```

```
    MSB = tmp & 0x8000;
    if (MSB != 0x0)
    {
     outb(SI, dataPort);
     outb(SCK | SI, dataPort);
    }
    else
    {
     outb(0x0, dataPort);
     outb(SCK, dataPort);
    }
    tmp = tmp << 1;
    outb(0x0, dataPort);
  }
  outb(CS, dataPort);
  return 0;
}
```

As you can see from this listing, programming of the MCP41010 chip is very simple and straightforward. The high byte of the `VCode` variable holds the command (0x11) while the data code (0 through 255) stored in the `Dn` variable determines the low byte. For this particular case I chose `Dn` equal to 100 since that corresponds to an output voltage of 1.98V at a given reference voltage of 5.08V. Note that the output voltage for a given circuit depends on the power source voltage applied to the endpoints of the potentiometer (5.08V in this case). Other possible binary codes of `Dn` for various output voltages are provided in commented lines just to support your own programming experimentation.

Programming of the output voltage requires writing a 16-bit value to the device and this is realized in the `for()` loop. The programming cycle starts when the CS signal goes from high to low thus enabling the write operation. After the final LSB has been written the program code raises the CS line thus ending the cycle.

You can replace the MCP41010 part with other similar devices. It is possible to use, for example, the MCP41050 device which offers a total resistance of 50K or the MCP41100 with a total resistance of 100K. This application could easily be expanded to use dual channel digital potentiometers MCP42XXX, although this evolution would require modifications in both hardware and software.

Use of digital potentiometers means we can easily build numerous projects having digitally controlled devices. We will discuss some of the applied practical uses of digital potentiometers later in this chapter.

So far we have talked about embedded projects dealing with signals incoming from analog sensors. The following project describes a temperature measurement system where we use a digital temperature sensor as input. The use of digital sensors has some extra benefits beyond analog counterparts.

When we take a digital sensor we have no need to build an analog signal chain since all analog conditioning circuits are implemented within the module; we need to provide only a control algorithm to implement the interface. However, digital sensors are presently less flexible compared to analog sensors when we try to configure resolution and accuracy.
Below I will describe a simple temperature measurement system using a popular digital sensor, the DS1722S chip.

Measuring temperature with the DS1722 digital sensor

The DS1722S sensor allows measuring temperatures from –55°C to +120°C with an accuracy of 2.0°C and requires **no** external components. Additionally, the DS1722S may be configured to operate with a resolution from 8 to 12 bits (1.0°C to 0.0625°C). Such a configurable resolution will be useful in applications where thermal runaway conditions must be detected most quickly. Data may be read from/written to the device via a SPI or 3-wire interface. In our project we will utilize the SPI interface. The schematic circuit of our system is shown in Fig.2.21:

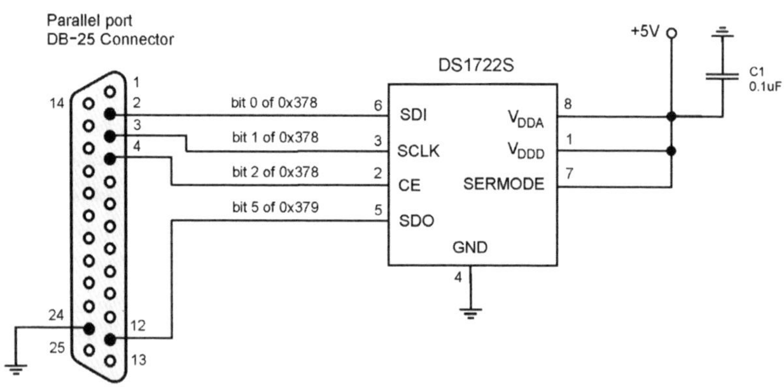

Fig.2.21

As you can see, the temperature sensor is attached to the parallel port of the PC via a 4-wire SPI compatible interface. This 4-wire mode is selected by connecting SERMODE (pin 7 of DS1722S) to the +5V supply. The device communicates with the PC via the following signal lines:

- CE (Chip Enable). This signal must be asserted high for communication to take place for either the SPI or 3-wire interface.
- SCLK (Serial Clock Input). This signal is used to synchronize data movement on the serial interface for either the SPI or 3-wire interface.

- SDO (Serial Data Output). When SPI communication is selected, the SDO pin is the serial data output for the SPI bus. When 3-wire communication is selected, this pin must be tied to the SDI pin (SDI and SDO pins function as a single I/O pin when tied together.)
- SDI (Serial Data Input). When SPI communication is selected, the SDI pin is the serial data input for the SPI bus. When 3-wire communication is selected, this pin must be tied to the SDO pin (SDI and SDO pins function as a single I/O pin when tied together.)

Before using the DS1722 sensor we need to configure it by writing an appropriate data byte to the configuration/status register containing the bits shown in Table 3.

Table 3. Configuration/status register's bits

bit 7	bit 6	bit 5	bit 4	bit 3	bit 2	bit 1	bit 0
1	1	1	1SHOT	R2	R1	R0	SD

SD is a 'shutdown bit'. If SD is "0", the DS1722 will continuously perform temperature conversions and store the last completed result in the temperature register. If SD is changed to "1", the conversion in progress will be completed and stored and then the device will revert to a low-power shutdown mode.The communication port remains active. The user has read/write access to the SD bit and the power up default is "1" (shutdown mode).

1SHOT bit means "One-shot temperature conversion" bit. If the SD bit is "1", a "1" written to the 1SHOT bit will cause the DS1722 to perform one temperature conversion and store the results in the temperature register at addresses 0x1 (LSB) and 0x2 (MSB). The bit will clear itself to "0" upon completion of the temperature conversion. The user has read/write access to the 1SHOT bit, although writes to this bit will be ignored if the SD bit is a "0". The power-up default of the one-shot bit is "0".
R0–R2 bits define thermometer resolution. In our project we will select an 8-bit resolution, so these bits will be set to 0. The user has read/write access to the R2, R1 and R0 bits and the power-up default state is R2 = "0", R1 = "0", and R0 = "1" that corresponds to 9-bit conversions.

The application can write the byte to the configuration/status register by using address 0x80. To read the data byte from the configuration/status register the application must use address 0x0. The temperature low byte (LSB) may be read at the address 0x1 and the high byte (MSB) may be read at address 0x2. In our project we will read the MSB only. Writing of the data byte to the sensor is presented in the timing diagram shown below (Fig.2.22):

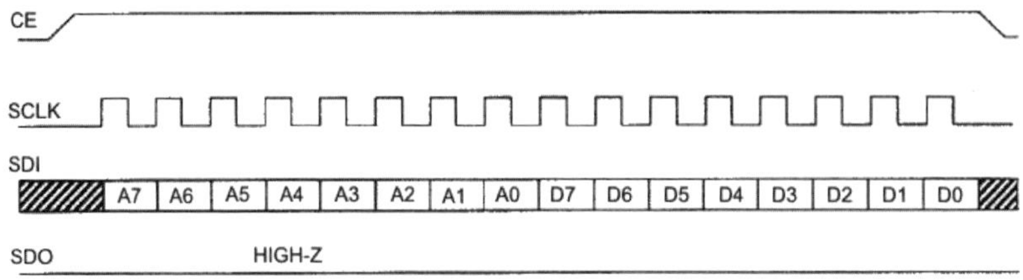

Fig.2.22

The address and data bytes are shifted MSB first into the serial data input (SDI) and out of the serial data output (SDO). Any transfer requires the address of the byte to specify a write or a read, followed by one or more bytes of data. Data is transferred out of the SDO for a read operation and into the SDI for a write operation. Each data transfer occurs when CE is high. Fig.2.23 shows the timing diagram when the data byte is read from the device:

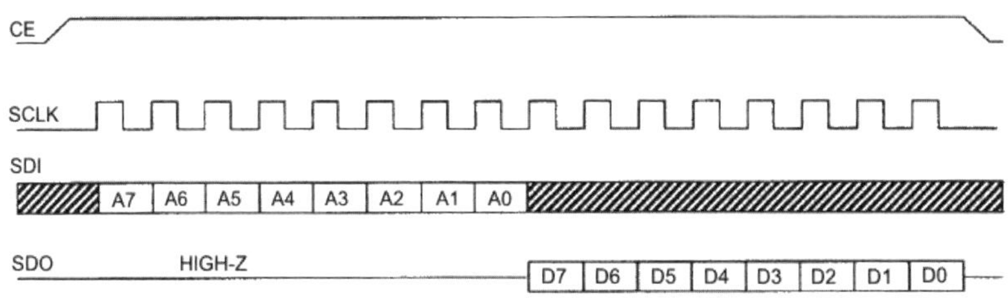

Fig.2.23

The detail information about interfacing and programming can be found in the appropriate datasheet on the DS1722 device. Now let's go to the software. The full C source code of the control application is shown below (Listing 2.10).

Listing 2.10

```
#include <stdio.h>
#include <unistd.h>
#include <stdlib.h>
#include <sys/io.h>

#define dataPort   0x378
#define statusPort 0x379
```

```
#define CE   0x4
#define SCLK 0x2


void dodelay(long del)
{
 while (del-- != 0);
}

void WriteByte(int addr, int data_byte)
{
 int i1, tmp;
 int MSB;
 int SDI;
 tmp = addr;

 for (i1 = 0; i1 < 8; i1++)
 {
  MSB = tmp & 0x80;
  MSB = MSB >> 7;
  SDI = MSB;
  outb(CE | SDI, dataPort);
  dodelay(10);
  outb(SCLK | CE | SDI, dataPort);
  dodelay(10);
  outb(CE | SDI, dataPort);
  dodelay(10);
  tmp = tmp << 1;
 }
 tmp = data_byte;
 for (i1 = 0; i1 < 8; i1++)
 {
  MSB = tmp & 0x80;
  MSB = MSB >> 7;
  SDI = MSB;
  outb(CE | SDI, dataPort);
  dodelay(10);
  outb(SCLK | CE | SDI, dataPort);
  dodelay(10);
```

```
   outb(CE | SDI, dataPort);
   dodelay(10);
   tmp = tmp << 1;
  }
}

int ReadByte(int addr)
{
 int i1, tmp;
 int MSB;
 int SDI, SDO;
 int result = 0x0;

 tmp = addr;
 outb(CE, dataPort);
 for (i1 = 0; i1 < 8; i1++)
 {
   MSB = tmp & 0x80;
   MSB = MSB >> 7;
   SDI = MSB;
   outb(CE | SDI, dataPort);
   dodelay(10);
   outb(SCLK | CE | SDI, dataPort);
   dodelay(10);
   outb(CE | SDI, dataPort);
   dodelay(10);
   tmp = tmp << 1;
  }
  for (i1 = 7;i1 >= 0;i1--)
  {
   outb(SCLK | CE, dataPort);
   dodelay(100);
   SDO = inb(statusPort);
   tmp = SDO & 0x20;
   if (tmp != 0x0)
   {
    tmp = 0x1;
    tmp = tmp << i1;
   }
   result = result | tmp;
```

```
    outb(CE, dataPort);
   }
  return result;
}

int ReadTemp(void)
{
  int total;
  outb(0x0, dataPort);
  WriteByte(0x80, 0x0); // Write config.
  outb(0x0, dataPort);
  sleep(1);
  outb(0x0, dataPort);
  total = ReadByte(0x2); // Reading MSB temp. reg. data
  outb(0x0, dataPort);
  return total;
}

int main(void)
{
  int Temp;
  iopl(3);

  Temp = ReadTemp();
  printf("Temperature = %d deg.Celsius\n", Temp);
  return 0;
}
```

Let's dig into this source code. To measure and calculate temperature we need some supplementary functions. One of them, `WriteByte()`, writes data byte specified by the `data_byte` parameter into the address pointed by the `addr` parameter. The `ReadByte()` function reads data byte from the address specified by the `addr` parameter and returns the result of the integer type. The `ReadTemp()` function processes the measurement and returns the binary code that represents an actual temperature as an integer value.
The `ReadTemp` function performs several operations. First, this function writes a configuration byte into the configuration/status register by executing the following statement:

```
WriteByte(0x80, 0x0); // Write config.
```

As you can see, the configuration byte is equal to 0x0. This means that the SD bit will be equal to "0" (continuous temperature conversion). In this case it doesn't matter what value the 1SHOT bit has, so

let's just set it to "0". Also we selected an 8-bit resolution by setting bits R0 through R2 to "0". After the configuration/status register has been configured the device executes a continuous temperature conversion. Since the measurement operation requires some finite time the program should wait by invoking the `dodelay()` function. After a given interval expires the application reads the temperature by calling the `ReadByte()` function. Since we are interested in knowing only the high byte we pass 0x2 as a parameter to this function.
We can save a source code in a DS1722_Measuring.c file and compile it:

```
$ gcc -o DS1722_Measuring DS1722_Measuring.c
```

Launching the application should be achieved using the following command:

```
$ sudo ./DS1722_Measuring
Temperature = 24 deg.Celsius
```

The application provides only one measurement and outputs the result onto a screen. You may modify the above code yourself to provide periodical measurements or build, for instance, a daemon which will collect results and save them in some file.That's all about this project.

Timing in Linux

Here, we will see how to build circuits where time dependencies are implemented. When we start talking about timing in desktop operating systems such as Windows or Linux we must understand that it is a difficult and complicated matter. Many times when we want to operate with tight time interval constraints and/or high frequencies we need special equipment such as expensive I/O expansion cards and device drivers for implementing some real-time tasks. Nevertheless, we can successfully implement many projects with demanding time requirements even with a home PC. There are methods for building such embedded projects.
First, you could take some multifunction chip which provides some real-time solution inside its chip. In this case you will not need to worry about implementing precision time circuits; all you must do is to program such a chip via some kind of interface (SPI, I^2C, etc.) to gain the functionality that you need. We have already discussed a similar solution when designing the wide-band oscillator based on the LTC6903 chip. As you recall, our control application had to configure that device via a SPI-compatible interface; all other jobs were performed by the chip. The benefit of using such an approach for designing your embedded system is obvious: you can concentrate on providing some simple hardware interface and programming the data transfer algorithm.

The second method to realize time-dependent embedded systems is to use simple circuitry with some low-cost and widely used components. You should entrust the main job in such projects to software. This approach works well for projects accepting timing resolutions to milliseconds. In these projects we may program time intervals within the capabilities of the operating system. This method allows practical realization of lots of useful time-constrained embedded systems, such as digital-to-analog converters, pulse-width modulators, etc. In the following sections we will discuss some useful and low-cost embedded projects where Linux interval timers are used.

Simple low-frequency oscillator

When we want to have a costless but precision square-wave oscillator operating up to 200-300Hz, we can accomplish this task entirely in software. The output signal may be taken from any of the data port pins. The only thing you need is to provide some buffering for an external load. The schematic circuit of such a software oscillator is shown in Fig.2.24:

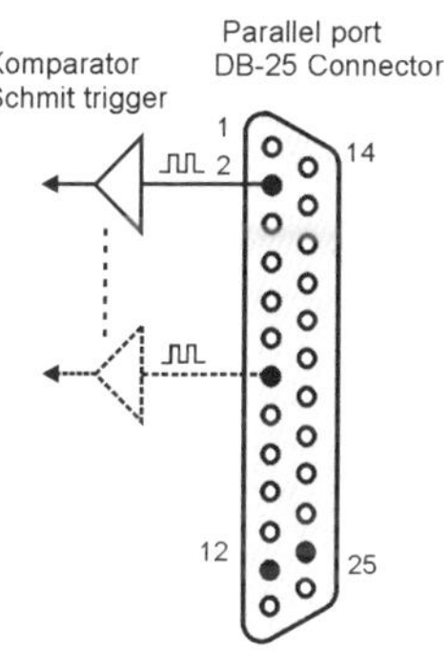

Fig.2.24

As you now know the parallel port's output buffers provide output currents of about 10-12 mA. When a load consumes a higher current you will need to connect hardware buffers between your parallel port hardware and the load. It is possible to use any kind of single-supply analog comparators with a high slew rate and TTL-compatible output or some Schmitt trigger chip from the TTL 74XX series. If you have a capacitive load you must ensure that your buffers can operate properly under such conditions.

The source code of our project is based on using an interval timer signaling with a frequency of 100Hz. The source code of our oscillator is shown below (Listing 2.11).

Listing 2.11

```
#include <stdio.h>
#include <signal.h>
#include <unistd.h>
#include <string.h>
#include <sys/time.h>
#include <sys/io.h>

#define dataPort 0x378
static char bRead;

void timer_handler(int signum)
{
  bRead = inb(dataPort);
  bRead = ~bRead;
  outb(bRead, dataPort);
}

int main(void)
{
 struct sigaction sa;
 struct itimerval timer;
 iopl(3);
 memset(&sa, 0, sizeof(sa));
 sa.sa_handler = &timer_handler;
 sigaction(SIGALRM, &sa, NULL);
 timer.it_value.tv_sec = 0;
 timer.it_value.tv_usec = 5000; // 5 ms interval before alarm
 timer.it_interval.tv_sec = 0;
 timer.it_interval.tv_usec = 5000;
 // time between signalling

 setitimer(ITIMER_REAL, &timer, NULL);
 while(1);
 return 0;
}
```

Let's discuss this source code. Here we use a timer event to implement our task. As all Linux (and UNIX) systems allow us to raise events caused by the timers we can apply this for our needs. To use an interval timer in Linux we have several methods at our disposal. One of them that provides a good timing is based on using the `setitimer()` function. In general, configuring the timer involves the following steps described later.
First, we need to configure the appropriate structures and set parameters for our timer. In our case we define structure `sa` of the `sigaction` type and `timer` of the `itimerval` type:

```
struct sigaction sa;
struct itimerval timer;
```

Then we determine the signal to be processed and point out the function which is called after a timer event has occurred. Two following statements do that:

```
sa.sa_handler = &timer_handler;
sigaction(SIGALRM, &sa, NULL);
```

In our case a `timer_handler` will toggle the output latches of the data port each 5 milliseconds. The timing intervals expressed in microseconds (μs) must be set for the `timer` structure:

```
timer.it_value.tv_sec = 0;
timer.it_value.tv_usec = 5000; // 5 ms interval before alarm
timer.it_interval.tv_sec = 0;
timer.it_interval.tv_usec = 5000;
```

Here `tv_usec` is set to 5ms (5000μs). The last thing we must do is to actually start the timer by executing the following statement:

```
setitimer(ITIMER_REAL, &timer, NULL);
```

The first parameter of the `setitimer()` function takes the `ITIMER_REAL` constant which means that the timer will work with real time intervals between events. The second parameter of this function points to the `timer` structure. After the timer has started the program enters an endless `while()` loop.
Even with low frequencies we can get great benefits when we apply some external circuitry. Once we operate with frequencies of 100 – 200 Hz we can build very useful embedded systems where these frequencies come in handy. First of all, we can generate a DC voltage controlled by frequency. This task can be realized using one of the circuits implementing a so-called "Frequency-To-Voltage" (F-To-V) conversion. A practical F-To-V circuit where the popular low-cost LM131/LM231 chip is used is shown below (Fig.2.25):

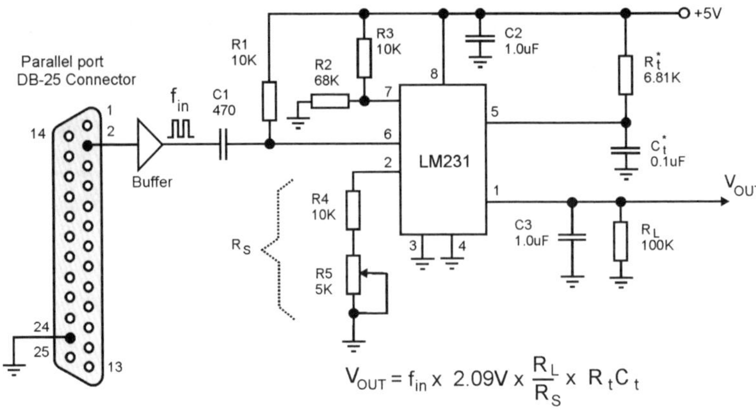

Fig.2.25

The "heart" of this circuit is a high-precision voltage-to-frequency converter device, the LM231, which we connect as a "frequency-to-voltage converter". The frequency fin from the parallel port goes to pin 6 of the converter via capacitor C1. As the formula implies (see Fig.2.25), the output voltage V_{OUT} of this device is linearly proportional to the input frequency fin. By changing fin we gain the particular output DC voltage V_{OUT} for handling and tuning miscellaneous circuits (robotics, sensors, etc.). In fact, we have achieved a digital-to-analog converter where the input signal is a frequency. You can easily customize this circuit by changing the values of Rt and Ct according to the recommendations in the manufacturer's datasheet on the LM231 chip.

Digitally controlled pulse width modulator

When we are considering how to get a DC signal from the frequency we may consider a pulse width modulation (PWM). The PWM is a most common method used in robotics and automatics for DC motor speed control. Pulse width modulation is the process of switching the power to a device on and off at a given frequency, with varying on and off times. These on and off times are referred to as "duty cycle". The diagram below (Fig.2.26) shows the waveforms of 25%, 50%, and 75% duty cycle signals.

As you can see from the diagram, a 25% duty cycle signal is ON for 25% of the wavelength and OFF for 75%, while a 75% duty cycle signal is ON for 75% and OFF for 25%. These signals are sent to the motor at a high enough frequency that the pulsing has no discernable effect on the motor. The end result of the PWM process is that the overall power sent to the motor can be adjusted from fully OFF (0% duty cycle) to fully ON (100% duty cycle) with good efficiency and stable control.

We can realize the simple PWM system using a very simple circuit and the program with source code described in Listing 11. Here is the basic circuit (Fig.2.27):

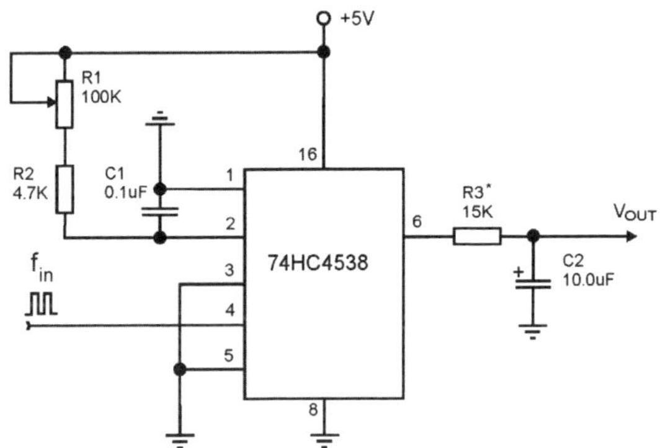

Fig.2.26

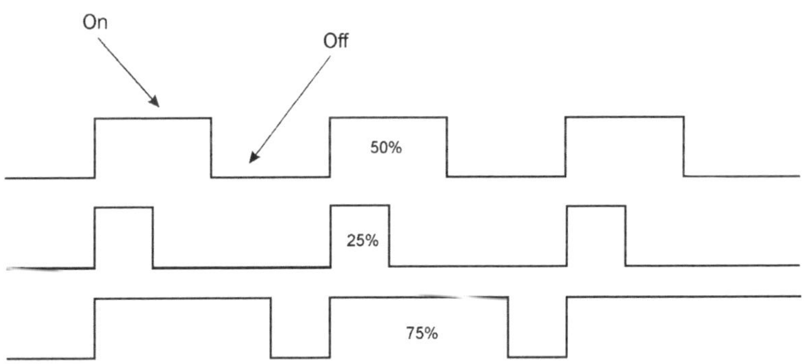

Fig.2.27

We use a 74HC/HCT4538 chip, that is a re-triggerable-re-settable precision monostable multivibrator. It can produce an output pulse with a width equal to 0.7 × (R1 + R2) × C1. The linear design techniques guarantee precise control of the output pulse width. Typical pulse width variation over temperature range is ± 0.2%. Also a Schmitt-trigger at the trigger inputs makes the circuit highly tolerant to slow rise and fall times. In our circuit the input signal goes to pin 4. The rising edge of

the input signal starts the multivibrator which outputs a pulse on pin 6. We can use this output pulse either for direct control of the DC motor via a power transistor or for generating the DC voltage using some basic a low-pass filter.
In our case, we obtain a DC voltage on the output of the simplest low-pass filter using the R3C2 pair. By shifting the wiper of R1 we can change the output pulse width over a wide range.
We can easily convert our circuit into a digitally controlled PWM source by replacing the variable resistor R1 with a digital potentiometer. In this case we could control the output voltage V_{OUT} programmatically from the parallel port via an SPI-compatible interface. The schematic circuit of such a digitally controlled pulse width modulator is shown below (Fig.2.28a).

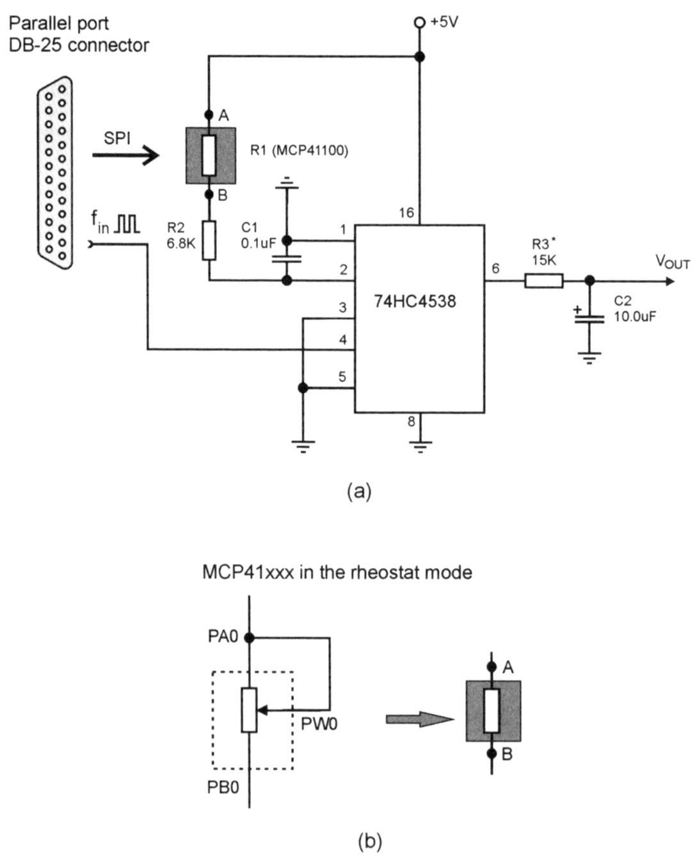

Fig.2.28

In this circuit we replaced a common discrete potentiometer R1 with the digital potentiometer MCP41100 that has a total resistance of 100K. We connected the MCP41100 according to the schematic in Fig.2.28b. Here the digital potentiometer operates in so-called "rheostat mode" where the device is used as a two-terminal resistive element. The unused terminal (PA0, in our case) is tied to the wiper (PW0). Reversing the polarity of the A and B terminals will not affect operation. Using the device in this mode allows control of the total resistance between the two nodes. The total measured resistance would be the least at code 00h, where the wiper is tied to the PB0 terminal. The resistance at this code is equal to the wiper resistance and may be about one hundred ohms for MCP41100 devices. The total resistance will increase as the digital code tends to 0xFF. The wiper will never directly connect to the PA0 terminal of the resistor stack. In the 00h state, the total resistance is the wiper resistance. To avoid damage to the internal wiper circuitry in this configuration, one should take care to ensure the current flow never exceeds 1 mA. In our case I increased up to the value of the limiting resistor R2 of 6.8K to guarantee that current flow will be less than 1mA.
For this circuit the input frequency fin passes from some output pin (2 through 9) of the data port of the parallel interface. With given values of R1, R2 and C1 the circuit could work at a frequency of about 100Hz providing a good linearity and a wide range of output voltage V_{OUT}. Recall, that we can easily provide the fin frequency of 100Hz programmatically using the source code provided in Listing 11. Also we can reuse the source code presented in Listing 9 to configure the digital potentiometer. By combining both listings (the software oscillator and configurable digital potentiometer) we may get the complete control program for the PWM circuit from Fig.2.28a.
You can find other popular PWM circuits on the Internet which can be customized to a digital control. For example, there are some popular PWM circuits based on the "immortal" 555 timer which may be easily modified so.

Using the I²C interface

Building measurement systems becomes significantly simplified when applying digital sensors. Digital sensors comprise all signal conditioning stages inside a single module chip and output the result as a binary code. Using such chips besides simplifying circuitry also saves electronic components, because digital sensors require few (if any) external components to provide their functionality. You may easily find many single-chip digital sensors to measure temperature, humidity, pressure, light, etc.

The digital code on the sensor's output may be transferred to the target system via one of the popular interfaces such as SPI, I²C, etc. This time we'll consider use of I²C. An I²C interface is a good solution when building a measurement system that must be controlled via a serial port of a PC. This is because this interface requires only two signal lines that the serial port can simply provide.

The other advantage of I^2C is that it allows multiple devices to be attached to the same bus. This feature greatly expands the capabilities of PC-based measurement systems where many devices need to be controlled. The simplified diagram of an I^2C bus with three devices attached is shown in Fig.2.29:

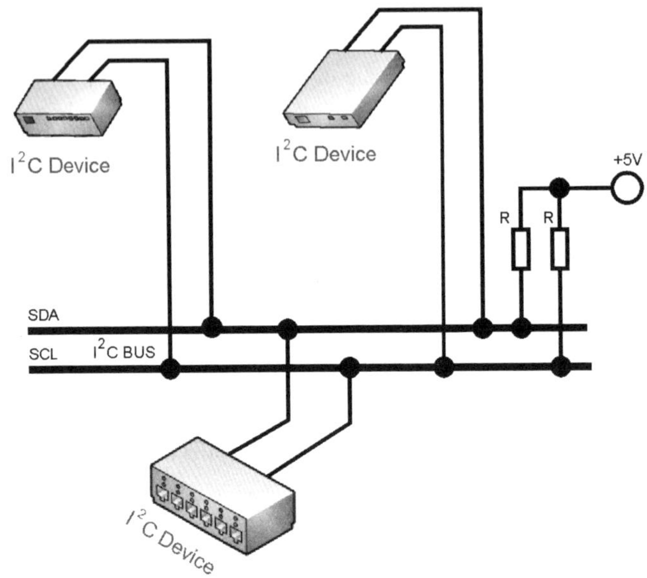

Fig.2.29

Communications over an I^2C bus require two wires called SCL and SDA. The SCL signal line is used to synchronize all data transfers via the I^2C bus, while SDA is the data line. All devices on the I^2C bus are connected to the SCL and SDA lines. Additionally, all devices must be attached to a third wire that is "ground" or "common wire" for a given circuitry. SCL and SDA operate as "open collector" or "open drain" signal lines, so we need two pull-up resistors between +5V power supply and the signal lines. Note, that the I^2C bus requires only two pull-up resistors, regardless of the number of devices attached. The "open collector" configuration allows driving of the line only when the device's output becomes low. The value of the resistors may vary from about 2KΩ to 30-40 KΩ. Greater values tend to slow the performance of the bus. The common value for pull-up resistors is 4.7K. Let's see how devices communicate on the I^2C bus.

Each device on the I²C bus can be either "master" or "slave". The "master" controls the bus by driving the SCL clock line. Other devices that are "slaves" respond to the "master". When working as the "master", a device can initiate a data transfer over the I²C bus. Usually, there is only single "master" on the I²C bus and multiple slaves. In embedded systems where I²C devices and a PC work together, the PC will usually be the "master", while all other devices function as "slaves". Note, that both "master" and "slave" can transfer data over the I²C bus, though that operation is always controlled by the "master". The simplified timing diagram of the I²C bus is shown in Fig.2.30.

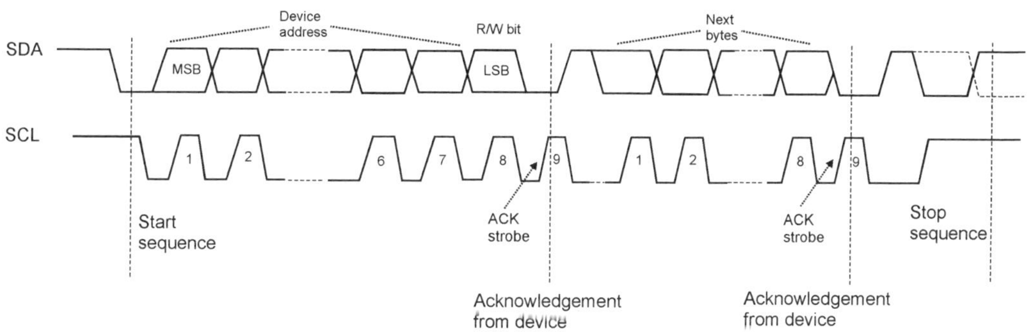

Fig.2.30

When working with an I²C bus, we should follow two rules. First, data transfer may be initiated only when the bus is not busy. Second, during data transfer, the data line must remain stable whenever the SCL clock line is high. Changes in the data line while the clock line is high will be interpreted as control signals. Thus, each data bit can be clocked only on a rising edge of the SCL pulse. The control signals are known as "start sequence" ("start condition") and "stop sequence" ("stop condition"). Each transaction with the "slave" begins with the "start sequence" and ends with the "stop sequence". Start and stop sequences are initiated by the "master". In order to issue the "start sequence" the master lowers the SDA line while the SCL remains high (see Fig.2.30).

The "stop sequence" is generated when the master raises the SDA line from low to high while the SCL is high. When data is being transferred, SDA must remain stable and not change whilst SCL is high. Various I²C devices may require different transactions, so the number of data bytes transferred between start and stop sequences is not limited and is determined by the "master".

Nevertheless, each transaction begins with the address byte which must contain the address of the "slave" device and the type of operation. Addresses of I^2C devices are either 7 or 10 bits. Most devices use 7 bit addresses that allow attachment of up to 128 devices on the I^2C bus. Besides the 7 bits of the address, the "master" always sends an 8th bit that indicates if the "master" is writing to / reading from the "slave"; if the bit is zero the "master" is writing to the "slave" , a value of "1" in this bit indicates that the "master" is reading from the "slave". The 7 bit address takes the higher 7 bits of the byte, while the Read/Write (R/W) bit is the LSB (Least Significant Bit) as shown in Fig.2.30.

Depending upon the state of the R/W bit two types of data transfer may be performed. When data is transferred from the "master" to the "slave", the address byte transmitted by the "master" contains the "slave's" address. Next follows a number of data bytes. The "slave" returns an acknowledge bit after each byte received.
The second case is when data is transferred from the "slave" to the "master". The address byte, as usual, contains the "slave's" address. The "slave" also returns an acknowledge bit. Next follows a number of data bytes transmitted by the "slave" to the "master". The "master" returns an acknowledge bit after all bytes has been received. At the end of the last received byte, a "Not Acknowledge" bit is returned.

The I^2C interface may be successfully applied for interfacing various devices to the parallel port as well as for expanding input/output signal lines when dealing with numerous digital signals. The following simple project illustrates how to expand the number of output lines of the parallel port using the PCF8574A I/O expander controlled through the I^2C bus. Below you can see the circuit (Fig.2.31):

This circuit is simple and consists of two chips. The buffer A1 forms TTLcompatible signals from the PC's parallel port to the PCF8574A I^2C I/O expander which is controlled via standard SCL and SDA lines. The control signals come from bits 01 of the data port of the parallel port. An 8-bit binary code (data byte) then appears on P0P7 pins of the expander. The LEDs attached to the P0P1 outputs will be triggered depending on the binary code received.

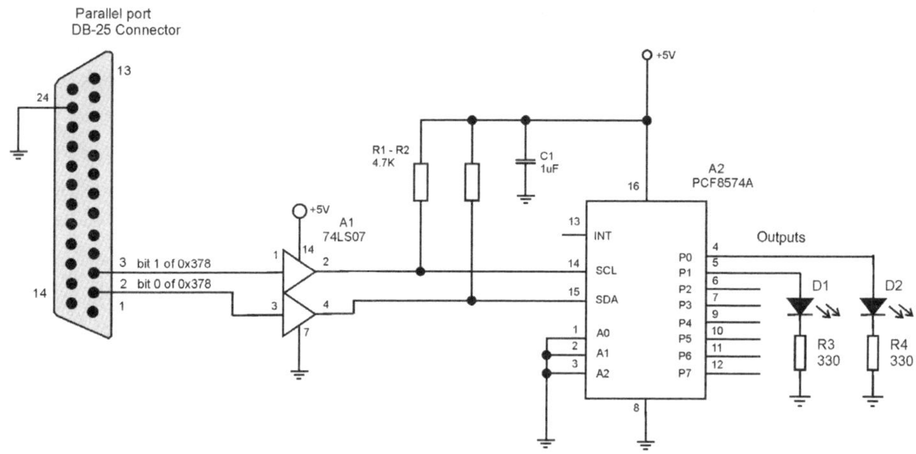

Fig.2.31

According to the I²C specifications, the data byte passes via the serial data (SDA) signal line, while the synchronization is provided by the serial clock (SCL) line. For a more complete understanding of how the timing diagram of I²C is implemented you may wish to refer to the datasheet of the PCF8574A chip. The A0-A2 address lines define the address of the device on the I²C bus. In a given case those lines are connected to "ground". Remember, that each device on the I²C bus must have a unique address. Anyway, when assigning the address you should consult the device's datasheet. That's all about hardware. Below you can see the C source code of the control application (Listing 2.12):

Listing 2.12

```
#include  <unistd.h>
#include  <stdio.h>
#include  <stdlib.h>
#include  <pthread.h>
#include  <sys/io.h>

#define  PORT1  0x378
#define  SCL  0x2
#define  SDA  0x1

//  D1  <->  SCL  --->  0x2
//  D0  <->  SDA  --->  0x1
```

```
char  dat;
int  cmd;
delay()
{
 int  cnt  =  30;
 while(--cnt != 0);
}
I2CStart()
{
 outb(0x0, PORT1);
 outb(SDA, PORT1);
 outb(SCL | SDA, PORT1);
 outb(SCL, PORT1);
}
I2CStop()
{
 outb(0x0,PORT1);
 outb(SCL, PORT1);
 outb(SCL | SDA, PORT1);// SDA goes high while SCL is high
}
WriteByte(char  b1)
{
 int  i1;
 for(i1 = 0; i1 < 8; i1++)
  {
   outb(0x0, PORT1); //  SCL goes low, SDA goes high
   if ((b1 & 0x80)!= 0)
   {
    outb(SDA, PORT1);
    outb(SCL | SDA, PORT1);
   }
  else
   outb(SCL, PORT1);

  b1 = b1 << 1; //  shift  left to the next bit
 }
 outb(0x0, PORT1);
 delay();  // Optional delay
 outb(SCL, PORT1);
```

```
}
main()
{
 iopl(3);
 int  cnt;
 for (cnt = 0; cnt < 4; cnt++)
 {
  I2CStart();
  dat = cnt;
  cmd = 0x70;
  WriteByte(cmd);
  WriteByte(dat);
  I2CStop();
  sleep(5);
 }
}
```

Here two variables, `cmd` and `dat`, are used. The `cmd` variable holds the code of the command that has to be sent to the I²C device in a first byte. The command is a combination of the address of the device and the code of the operation (read/write). In our case, the code of the command will be equal to 0x70. The `dat` variable contains the data byte that appears on P0-P7 outputs after a write cycle is completed.

Writing data via the I²C bus for the PCF8574A device takes four steps: issuing the "start sequence", writing the first byte which contains the command, writing the data byte that must be on the P0-P7 outputs and issuing the "stop sequence". These steps will be executed inside the main program that invokes the appropriate functions in the right order. The `I2CStart` function generates the "start sequence". The `WriteByte(cmd)` function passes the first byte (command, `cmd` variable) to the PCF8574A device. Then the `WriteByte(dat)` function passes the data byte which is outputted to the P0-P7 pins. Finally, the `I2CStop` function generates the stop sequence which ends a data transfer via the I²C interface.
The design may be easily improved upon by increasing the number of outputs up to 16 after replacing the PCF8574A chip using a PCF8575C. Also, there are other I²C expanders that may be used in this design, for instance, MCP23017 by Microchip, etc. When you make such replacements you should also change the schematic circuit in Fig.2.31 in an appropriate way and rewrite the source code of the application according to the timing diagram of the selected chip.

Recommendations and Conclusion

We discussed some useful projects achievable by using a standard parallel port of the PC and some popular software. All projects will also work with PCI expansion boards equipped with the parallel interface as well. When you use an expansion board, replace addresses 0x378 and 0x379 with the values assigned to the PCI card by your system. When designing projects you should care about proper layout of your circuits.

When laying out circuits for your embedded systems some precautions must be implemented. Place bypass capacitors as close as possible to device power pins. A bypass capacitor value requirement may vary depending on the chip; therefore you must consult a datasheet on any particular part. Digital and analog traces should be separated as much as possible on the board, with no traces running underneath the device or the bypass capacitor. Extra precautions should be taken to keep traces with high-frequency signals (such as clock lines) as far as possible from analog traces. Use of an analog ground plane is recommended in order to keep the ground potential the same for all devices on the board.

Chapter 3. Home projects with a serial port

This chapter continues the discussion concerning home project designs. Here we will see how to build various embedded systems using a serial port (interface) of the PC. The serial interface that is also known as RS-232 is a standard interface used for data transfer between various devices, such as modems, PCs and communication devices. Many developers and hobbyists are familiar with this primary function of RS - 232. However, serial interfacing may be successfully applied in designing embedded projects able to perform various data acquisition and control functions. Many such projects can be developed, even at home, with minimal costs and effort. The serial port protocol - together with data transfer systems - allows building and integration of home or laboratory embedded systems with complex post-processing data processing algorithms.
Since we are going to discuss the serial port hardware and software, you should make yourself familiar at least with RS-232 basics. Wewill not discuss RS-232 in detail; instead we will cover the practical aspects of using RS-232 features in real projects. The great capabilities for data processing are provided by the popular MATLAB® development environment from The Mathworks, Inc. Using MATLAB developers, researchers and hobbyists can realize many complex projects, including the design of embedded systems based on using PC peripherals (serial port, sound card, etc.). We will consider serial port based examples developed especially to work in the MATLAB environment.

Serial port basics

The heart of the serial interface is a special register-based digital circuit called a UART. The term "UART" stands for Universal Asynchronous Receiver / Transmitter. Most motherboards used in desktop PCs and notebooks have UARTs integrated in their chipsets and similarly expansion cards often have a built-in serial interface. Despite the variety of practical realizations of serial interfaces most of them use modern UART chips compatible with a classic 8250 UART device. The most popular and widely used derivatives presented on the market are 16550/650/750/etc. chips. All of these share backward compatibility and have common basic hardware registers and signal lines which will be useful for our home projects.

Although modern UARTs are complicated devices with complex architectures, their operation is simple enough from the developer's point of view. When dealing with a UART, we should be aware of two things:

- How to program it. Being a modern device, a UART is configured and controlled using some hardware registers (ports), so we can gain full control over our UART by accessing its hardware registers programmatically;

- How to connect it to external devices. The following simplified diagram shown in Fig.3.1 illustrates interfacing a UART to external circuits:

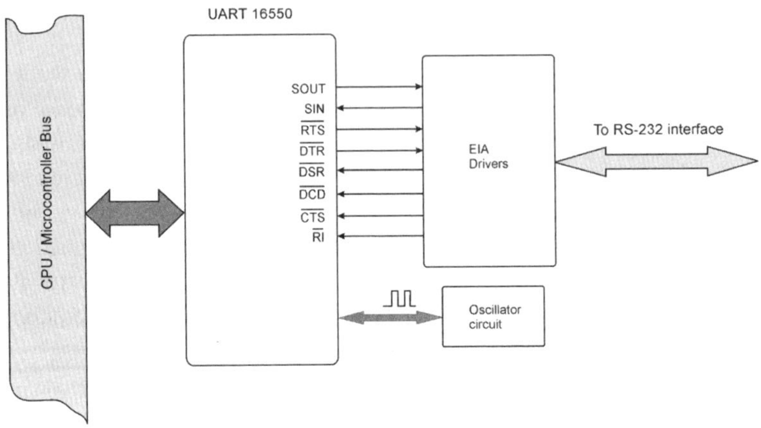

Fig.3.1

As you can see, the UART may be controlled from a microcontroller or microprocessor as shown on the left side of Fig.3.1. The baud rate is configured using an oscillator circuit and software division of a clock. Configuring modes of operations is implemented by writing particular values into internal hardware registers which are responsible for the UART's functioning. Each standard UART may be accessed through 7 hardware registers. As usual, the addresses of these registers are relative to some "base" address. In the case of a built-in UART,as will be found on most motherboards, this address will usually be 0x3F8; expansion cards may have other base addresses depending on how its' devices are configured by a PnP subsystem of the Linux operating system.

A UART interacts with external devices using several output signals (eg. SOUT, RTS and DTR) and several input signals (eg. SIN, DSR, DCD, CTS and RI) which pass through RS-232 line-driver devices to form standard voltage levels for the serial interface.

An end user can see all these signals on the DB25 connector on the back of the PC. Except SOUT which appears as TD and SIN which is assigned RD, all other signals keep their original names at the DB25. SIN and SOUT signals are data lines used for data transfer between communication devices. Both these lines may be used to create a so-called "null-modem" link which allows transferring of data between two communication devices without any additional signals. With this configuration the SOUT (TD) signal line of the first device is connected to SIN (RD) of the second device and vice versa, SIN (RD) of the first device is connected to SOUT (TD) of the second device. The "null-modem" configuration is often used to connect a microcontroller development board to the PC. The "null-modem" configuration is shown in Fig.3.2:

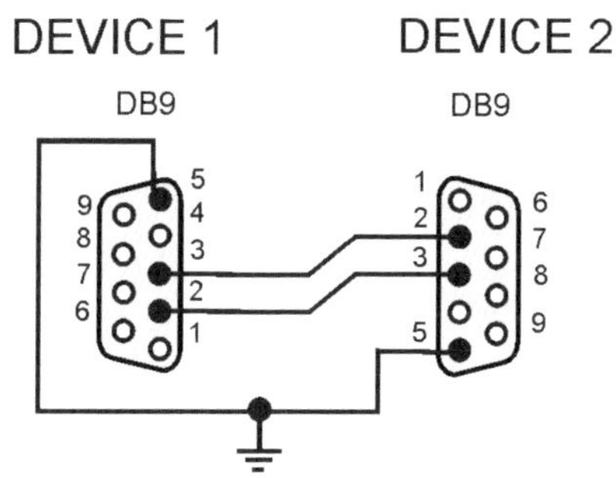

Fig.3.2

Here the TD pin of DEVICE 1 is connected to the RD pin of DEVICE 2 and TD of DEVICE 2 is attached to the RD pin of DEVICE 1. This configuration allows bidirectional data flow between both devices with minimal wiring. Such an interface is very frequently used when transferring data between PCs and microcontrollers. The only requirement for this configuration is that both devices should operate at the same frequencies.

From a first point of view it seems that the developer can't squeeze anything else from SIN/SOUT signals besides their basic functions, but soon we will see how these data transfer lines may be successfully applied to control embedded devices.

Besides data lines the serial interface comprises more signal lines known as control lines. We can use two output lines (RTS, DTR) and two input lines (CTS and DTR) in an embedded design. All those signals were initially introduced for synchronization ("handshaking") of data transfer, but may come in handy when designing miscellaneous embedded projects. Let's see how to operate with these signals in our control applications.

Each time we want to manipulate one of the RTS, DTR, CTS or DTR signals we need to access two hardware registers, Modem Control Register (MCR) and Modem Status Register (MSR). The MCR register (port) contains bits through which the program can set/clear RTS (Request-To-Send) and DTR (Data-Terminal-Ready) signals. MCR has the address that is calculated as "base + 4". For the standard UART where the base hardware address is 0x3F8 the MCR register will have the address 0x3FC. The MCR port is accessible for reading/writing operations. The RTS line may be placed in active state (low voltage level) by setting bit 1 of the MCR. Again, deactivating the RTS line (raising a voltage level) occurs after bit 1 has been cleared. As you can see, the level of the RTS signal will be inverted in relation to the state of the RTS bit in MCR.

The DTR signal line is also controlled via MCR. To handle DTR you have to set/clear bit 0 of the MCR register. As in the case of RTS, when the DTR bit is low the DTR signal line is high and vice versa.

Other signal lines, CTS (Clear-To-Send) and DSR (Data-Set-Ready), are inputs and may be involved in reading data from external devices. Both lines may be read via the Modem Status Register (MSR) having an address equal to "base + 6". Bit 4 of MSR reflects the state of the CTS line and bit 5 does the same for the DSR line. Both bits are complements of CTS and DSR signals. Thus, for example, when bit 4 of MSR is set the CTS line goes low; clearing bit 4 results in raising the voltage level on the CTS line.

Let's see how serial interfaces are identified in the Linux.

Usually the serial interfaces in Linux are configured as `/dev/ttyS*` devices.

To discover what serial interfaces we have, enter the `dmesg` command together with a `grep` command in a terminal window. Below you can see all the available serial interfaces on my desktop PC :

```
$ dmesg | grep ttyS*
[ 0.001365] console [tty0] enabled
[ 0.963104] serial8250: ttyS0 at I/O 0x3f8 (irq = 4) is a 16550A
[ 0.963529] 00:06: ttyS0 at I/O 0x3f8 (irq = 4) is a 16550A
[ 5.983375] 0000:01:05.0: ttyS1 at I/O 0x9000 (irq = 21) is a 16550A
[ 5.983538] 0000:01:05.0: ttyS2 at I/O 0x9400 (irq = 21) is a 16550A
```

From this output you can see that three serial interfaces, `ttyS0` through `ttyS2` are determined in the system. The first, `ttyS0`, is the standard serial port built-in on the motherboard, while the others (`/dev/ttyS1` and `/dev/ttyS2`) belong to a PCI expansion board. This output brings up only a basic hardware configuration; more detailed information may be retrieved by entering the `setserial` command. This command helps us to set and/or report the configuration information associated with a serial port. This information includes the address of the I/O port and the IRQ number of a particular serial port in use, and whether or not the break key should be interpreted as the Secure Attention Key, and so on. Getting information about all serial ports installed, for instance, may be retrieved by the following command:

```
$ setserial -ag /dev/ttyS[012]
```

Below you can see the output produced using the above command on my own Linux system:

```
$ setserial -ag /dev/ttyS[012]
/dev/ttyS0, Line 0, UART: 16550A, Port: 0x03f8, IRQ: 4
  Baud_base: 115200, close_delay: 50, divisor: 0
  closing_wait: 3000
  Flags: spd_normal skip_test

/dev/ttyS1, Line 1, UART: 16550A, Port: 0x9000, IRQ: 21
  Baud_base: 115200, close_delay: 50, divisor: 0
  closing_wait: 3000
  Flags: spd_normal skip_test
```

```
/dev/ttyS2, Line 2, UART: 16550A, Port: 0x9400, IRQ: 21
  Baud_base: 115200, close_delay: 50, divisor: 0
  closing_wait: 3000
  Flags: spd_normal skip_test
```

Here we obtain more detailed information about serial interfaces. Specifically, we can see what UART type is used or what baud rate is set for the particular port. The `setserial` command is also used to configure parameters of the particular port. The following command, for example, allows setting the baud rate of `/dev/ttyS1` to the value of 9600 baud:

```
$ sudo setserial /dev/ttyS1 baud_base 9600
```

Again, to check the settings just performed we can launch the following command:

```
$ setserial -ag /dev/ttyS1
/dev/ttyS1, Line 1, UART: 16550A, Port: 0x9000, IRQ: 21
  Baud_base: 9600, close_delay: 50, divisor: 0
  closing_wait: 3000
  Flags: spd_normal skip_test
```

The `setserial` command takes many other parameters which you can set when configuring the serial port. For further details concerning `setserial` the Internet offers lots of reference manuals on such a given topic.
For all supported UNIX platforms you may also use the `stty` command. This command allows displaying or configuring serial port information. For example, to display serial port properties for `/dev/ttyS0` you may enter:

```
$ stty -a < /dev/ttyS0
```

To configure the baud rate to 9600 bits per second you may type in:

```
$ stty speed 9600  < /dev/ttyS0 > /dev/ttyS0
```

Before using a serial port it is worth a simple check that its working exactly as you would expect it to. The simplest way to do that is to send some data out via the TD pin and receive the same data back in via the RD pin. Naturally, this doesn't guarantee fully 100% operability of the serial interface, but allows good estimating of the serial port functionality on the whole. In this case we use a so-called "loopback interface" arrangement which is shown in Fig.3.3.

Male RS-232 DB9 Connector

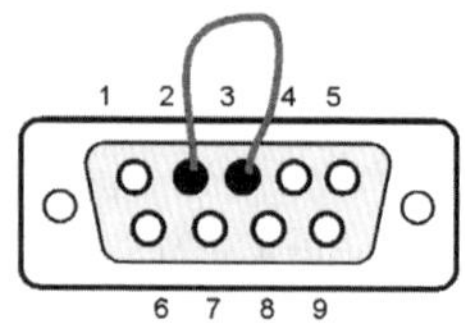

1 Carrier Detect (CD)
2 Received Data (RD)
3 Transmitted Data (TD)
4 Data Terminal Ready (DTR)
5 Signal ground
6 Data Set Ready (DSR)
7 Request To Send (RTS)
8 Clear To Send (CTS)
9 Ring Indicator (RI)

Fig.3.3

After TD and RD pins are shorted together in this way you may send, for example, a single byte to the particular interface and try to receive the same byte back.
Suppose we want to check the `/dev/ttyS1` serial interface. First, open the terminal window and type in the following command:

```
$ cat < /dev/ttyS1
```

This command while executing waits for some data to arrive at the RD pin. Second, open the next terminal window and type in the following command:

```
$ echo Test string has been received > /dev/ttyS1
```

After pressing Enter the text string will be sent via the TD pin. If `/dev/ttyS1` operates properly, you will see the received string in the first window.
In my system I saw the following result (Fig.3.4 – Fig.3.5).

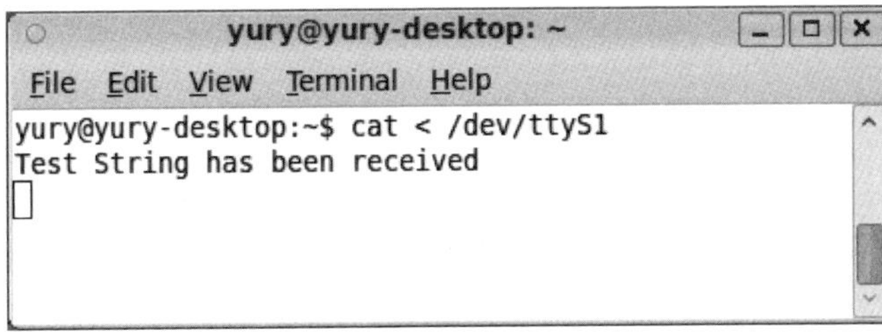

Fig.3.4

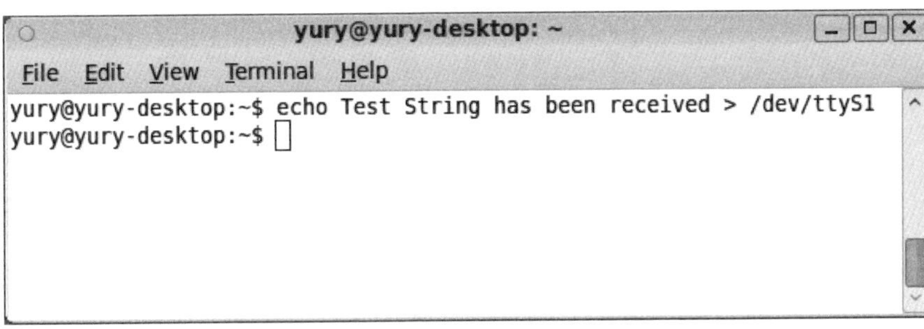

Fig.3.5

Now let's see how we can use the serial port data lines in embedded projects.

Using data lines in embedded projects

Let's see how we can apply the TD line to implement simple home automation tasks. Recall that the data stream through TD line is a sequence of bits. Each bit in this sequence is represented as a voltage swing that ranges from, nominally -12V to +12V. We may apply such a bit sequence to switch on/off external devices attached to the RS-232 interface. Implementing this task requires some simple additional circuitry which may be easily assembled and tested.
Below (Fig.3.6) you can see a simple demo circuit which allows switching on a DC load (a LED, in our case) when a byte is sent through the TD pin.

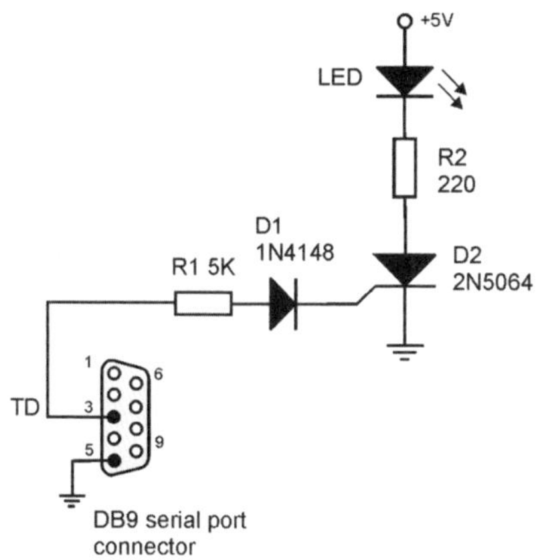

Fig.3.6

Here the TD pin is connected to the gate of SCR 2N5064 (D2) through the resistor R1 and diode D1. When a single byte is being transmitted through TD, a bit sequence turns on the SCR device 2N5064 and the LED device begins lighting. As the pulses on TD range from -12V to +12V, we need the D1 diode to pass only the unipolar signal which switches on the SCR device through its gate.

Suppose, we have our circuit attached to the `/dev/ttyS0` device. In this case we can turn on the LED by writing any character to the given serial port:

```
$ echo t >/dev/ttyS0
```

This circuit may be easily modified to handle AC loads (we will discuss this later).

The circuit shown in Fig.3.6 can be improved to switch on/off the DC load each time the next byte is sent via TD. The following circuit (Fig.3.7) illustrates this concept:

Here the bipolar signal in the range of -12V to +12V goes to the base of transistor Q1 operating in a switch mode. Q1 forms the unipolar series of pulses with magnitude within the range of 0 to about +5V that corresponds to TTL logic signals. The TTL pulse on the collector of Q1 goes to the input pin 2 of a LM555 timer configured as a monostable multivibrator. Upon the falling edge of the pulse on pin 2 of the timer, the LM555 generates a single pulse on output pin 3. The duration of this pulse will be about 0.7 × R3 × C1 and should be chosen to exceed the total interval taken by UART to transfer a single byte. In this case each byte being transmitted via TD will cause a single pulse on the A1 output.

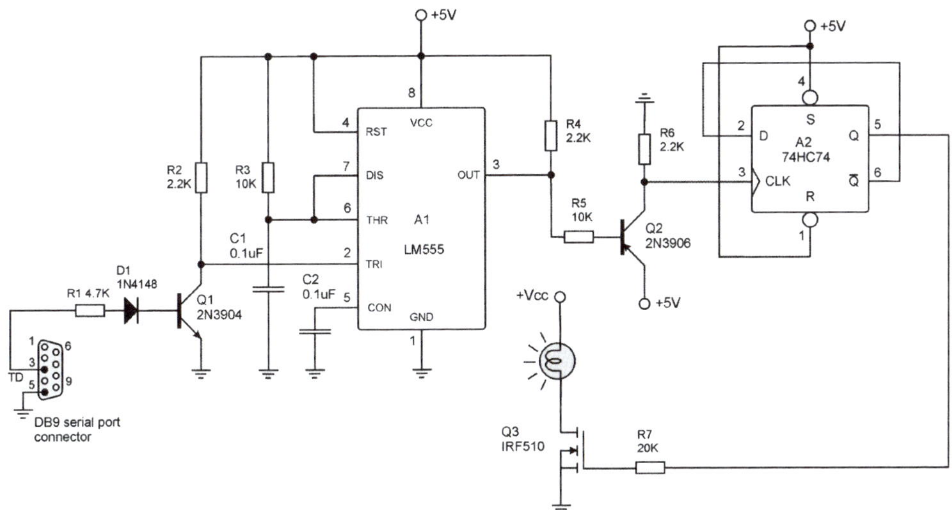

Fig.3.7

The values of R3 and C1 may be chosen over a wide range. With the given values of R3 and C1 the pulse duration will be about 1mS. The next stage after the 555 timer is a switch (Q2) which drives the D-type flip-flop, type 74HC74. The rising edge on the CLK input toggles both outputs of A2. As the gate of the power MOSFET IRF510 is connected to the Q output, the external load attached to the drain of Q3 is also toggled. Note that this circuit works with a DC load (bulb, in our case) attached to the drain of Q3.

Toggling the load is implemented by sending a single byte out of the serial port. For example, if a given circuit is wired to the `/dev/ttyS0` serial interface you can use the following command:

```
$ echo t >/dev/ttyS0
```

to toggle the load.

This circuit can be modified so as to control various AC loads at home. You would consider, for example, the following circuit (Fig.3.8):

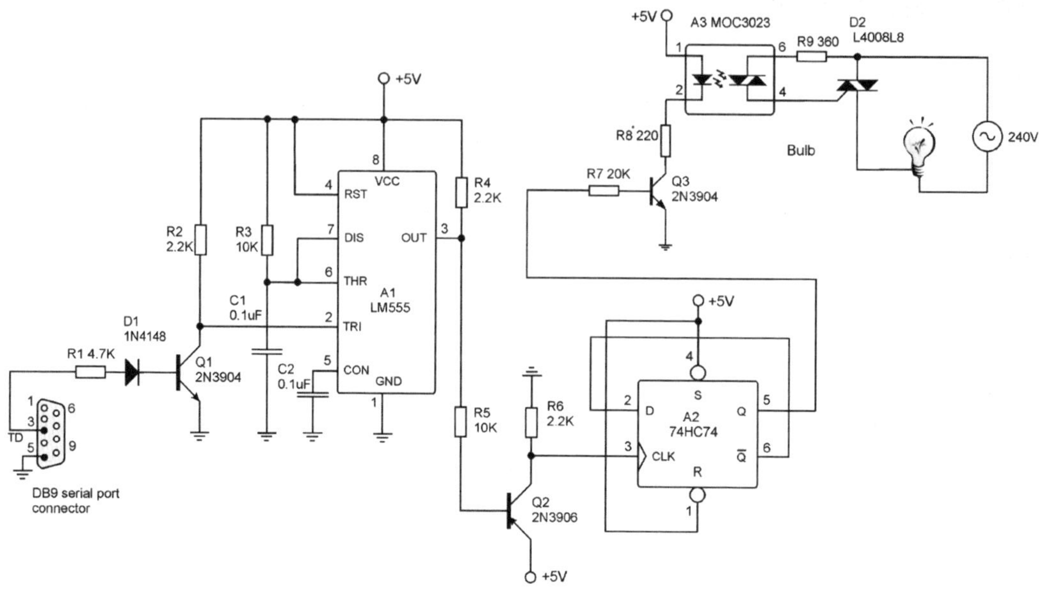

Fig.3.8

Compared to circuit shown in Fig.3.7, the Q output of the flip-flop toggles the switch Q3 which, in turn, activates the optotriac MOC3023 subcircuit (A3 in Fig.3.8). The high-voltage side with an AC bulb will be driven by the optotriac A3 through the gate of the power triac D2.

Disclaimer! Caution! If you do not possess experience in designing mains powered AC power circuits, do not attempt to connect your circuits to mains voltages. There is a real danger of injury or death from electrocution or burnse! The only way is the safe way - if you still want to work with AC power circuits - and that is to call a specialist to assemble and check the circuit for you. Remember that the device assembled must be put into an enclosed box fabricated from a suitably insulating material.

Embedded projects such as just discussed demonstrate use of the TD signal to control either DC or AC external loads. If, however, we want to receive some signal and capture it we can use the RD signal line; we will consider how to do that in the next section where a simple security system is discussed.

Home security system

First of all let's look at the hardware. It consists of the Passive Infrared Sensor (PIR) and only several more components. The PIR-circuit is attached to the serial port /dev/ttyS0 via the optotriac MOC3063. First of all let's see in brief what a PIR-sensor is and how it works.
A passive Infrared Sensor is an advanced version of a motion sensor. PIR sensors can identify the presence of people by detecting the infrared radiation that people emit, which stems from body heat. The PIR sensor is completely passive and reacts only to drastic changes in levels of infrared radiation emitted in their surveillance area, typically caused by the movement of a person in a room. This prevents false alarms when the floor cools at night, for instance. These sensors are designed so that they don't react to pets moving around in their surveillance area.
In our project we use the PIR detector module KC7783. This is a fully assembled and tested miniature PIR module (KC7783 with KC778B IC) with only 3 connections: Power, Ground and Output. The module can be powered from +4 – +12V DC power supply (in our project we use the standard TTL-compatible 5V power source). In this module RE200B PIR sensor is used. The output signal is an active high pulse of approximately 0.5 seconds minimum which remains active as long as there is motion. The device can "sense" such motion withrange up to 10 feet. The size of this module is35mm x 25mm; it can be easily mounted on the wall. In fact, you can freely apply almost any PIR-sensor manufactured for most generate TTL-compatible signals.
Besides the PIR sensor module there are also some additional electronic components that are needed to interface the PIR sensor module to the serial port. The full schematic circuit of our device is shown in Fig.3.9:

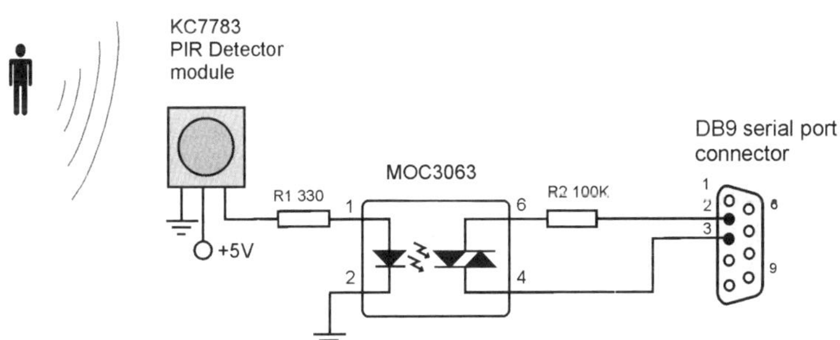

Fig.3.9

This application operates in the following way. The control program running on the PC continuously sends data through the TD output data line (pin 3 of the DB-9 connector) of the serial port and tries to receive the same data on the RD line (pin 2 of DB-9). If the PIR sensor is not being excited it yields a low-level voltage on its output, so the LED of the optotriac is off. That, in turn, holds the triac closed, therefore nothing goes from TD to RD because of a very high resistance between TD and RD.

As the program doesn't receive any data back, this means that nobody was caught inside the scanned area. If a human body crosses the active area covered by the sensor, the PIR detector activates and generates a high TTL-compatible level on its output. This signal switches on the LED which turns on its opto-triac .. From this point the relatively low resistance of R2 will not prevent the signal from passing from TD to RD and the control program receives the signal it sent itself. Once that happens, the program considers that an intrusion occurred and acts appropriately (turns on alarm sound, locks the area, etc.). The RS-232 interface allows wiring distance uo to 12-15m, so you can place the PIR-sensor far enough from your PC. This circuit comprises a widely-used optotriac type MOC3063 which may be driven by a low current (LED trigger current > 5mA). As many modern PIR-sensors provide low output current, we can directly connect a given optotriac to the sensor using a single resistor. However, if you cannot obtain an optotriac with a low-level LED trigger current you can slightly modify the circuit shown in Fig.3.9 so to accommodate optocouplers with a higher LED trigger current.

All you need do is to introduce some buffer between a PIR-detector's output and an optotriac. One possible solution is shown below (Fig.3.10):

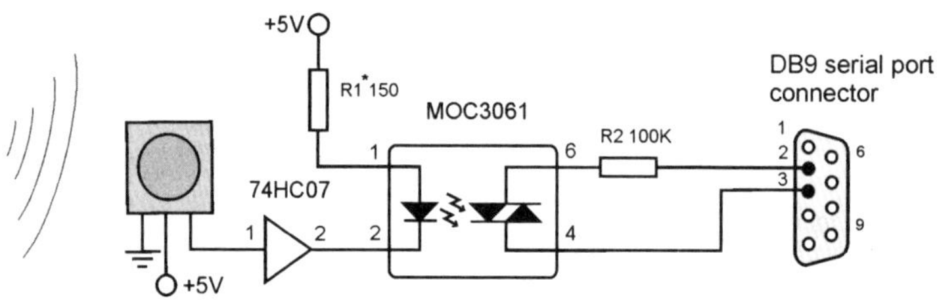

Fig.3.10

Here the buffer, type 74HC07, with an "open collector" (OC) output provides enough current (≤ 20mA) through a LED to turn on the optotriac, type MOC3061. In order to control the circuits shown in Fig.3.9–3.10 you either write a special program or simply use a couple of commands entered in a terminal window. Suppose, we attached our circuit to the `/dev/ttyS0` serial interface. Then we need to open a terminal window and enter the command:

```
$ cat < /dev/ttyS0
```

In the other terminal window the following command would be entered:

```
$ echo Alarm! > /dev/ttyS0
```

Press the 'Enter' key and look what happens. If a PIR-sensor has been activated, the text "Alarm!" appears in the terminal window where a `cat` ... command is running. You may put the `cat` ... command into a background by adding "&" at the end the command. In this case you must remember the process identifier (PID) of the given process, so that you can terminate the background process when needed.
Entering the following command will output it's PID onto the screen:

```
$ cat < /dev/ttyS0&
[3] 2563
```

As you can see, the PID is equal to 2563, so you can terminate the process having this PID by using the following command:

```
$ sudo kill 9 2563
[3]    Terminated     cat < /dev/ttyS0
```

Besides putting the `cat`... command into a background, you may also schedule checking of the PIR-detector by periodically polling its state. This requires development of a simple BASH-script, which will execute the `echo ... command` after some predefined interval expires
Below you can see the source code of such a demo script where we check the state of this security system every 5 seconds:

```
#!/bin/bash
while
 true
do
  echo ALARM! > /dev/ttyS0
  sleep 5
done
```

Here the "ALARM!" string is sent every 5 seconds to the /dev/ttyS0 device inside the endless `do...` loop. Reception of this string must be checked using the `cat < /dev/ttyS0&` command , running in a background. You must save this source code in a file and make this file executable. For example, after we have saved the source in the file named as `program1` we need to set an appropriate permission using a `chmod` command:

```
$ chmod +x program1
```

After that we can launch program1 by entering the command:

```
$ ./program1
```

It is assumed that we are in the directory where program1 resides. Note that program1 could be started in a background just like the cat... command! As you can see, we have created a simple but actually quite novel security system.
To extend the capabilities of our security system we can improve the software by writing "daemon" applications. Daemons can accomplish additional actions that may be difficult to implement using single commands or even BASH scripts.. The C source codes of simplified versions of both daemons are presented below (Listing 3.1 and Listing 3.2):

Listing 3.1

```
#include <stdio.h>
#include <unistd.h>
#include <stdlib.h>

int main(void)
{
 pid_t PID;
 PID = fork();
 if (PID == 0)
  exit(0);

 setsid();
 printf("Write Daemon started...\n");
 while (1)
 {
  system("echo ALARM > /dev/ttyS0");
  sleep(5);
 }
return 0;
}
```

Listing 3.2

```
#include <stdio.h>
#include <unistd.h>
#include <stdlib.h>

int main(void)
{
 pid_t PID;
 PID = fork();
 if (PID == 0)
  exit(0);

 setsid();
 printf("Read Daemon started...\n");
 while (1)
 {
  system("cat < /dev/ttyS0");
   < other statements >
}
return 0;
}
```

To terminate daemons you should previously determine their PIDs using the `ps -ef` command and kill the processes with the `kill` command, as discussed above.

As you can see there are various methods of realizing control software for a given security system. Nevertheless, we can also improve the hardware of this system. Below you can see a schematic circuit with which it is possible to cover an additional area by placing extra PIR-sensors in the room to be scanned (Fig.3.11):

In this circuit we use 4-input OR gate, type 74HC4072, to collect signals from miscellaneous sensors scattered over the area of interest. The optotriac, type MOC3063, is turned on when the signal from at least one sensor has signaled on one of the four inputs of the OR- gate.

That's all about this embedded project. As you can see, we innovated a solution based on a non-standard use of the data lines (TD and RD). I encourage you to think about other projects that may be developed using this approach.

Let's consider other capabilities the serial interface offers. Recall that the serial interface has several control lines (RTS, DTR, CTS and DSR) whose purpose is usually data flow control (handshaking).

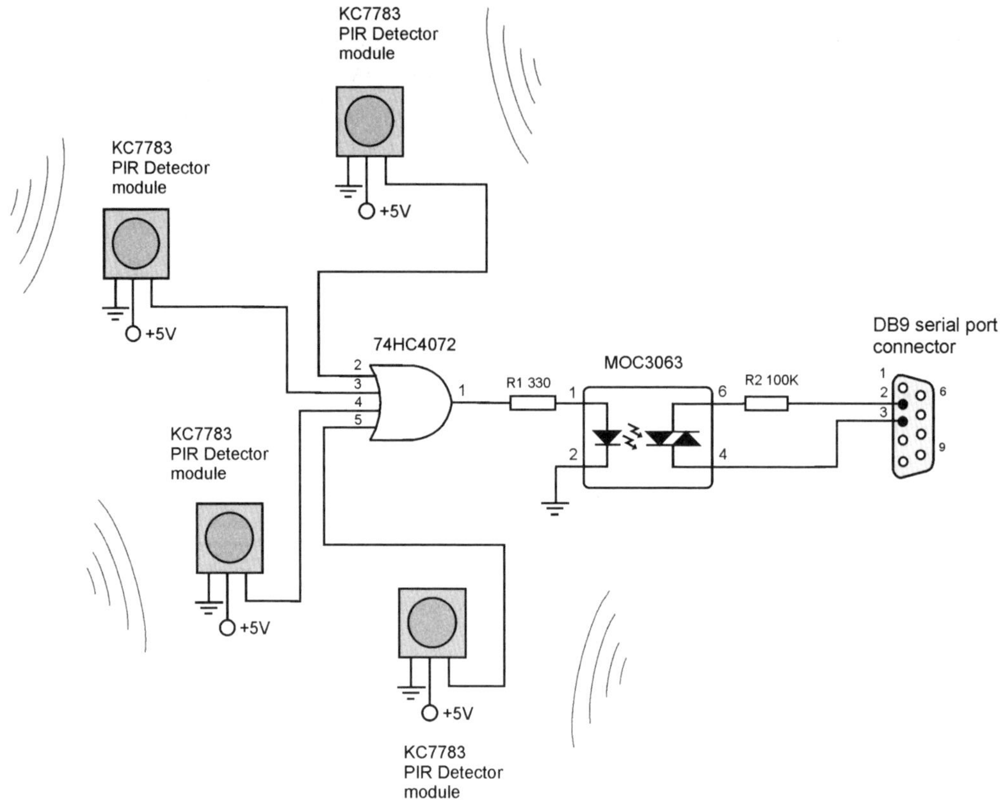

Fig.3.11

But nothing forbids us to use those lines for our own purposes. Of those lines two (RTS and DTR) are digital outputs that can be freely set/cleared by software when needed. The other two (CTS and DSR) are digital inputs which can be read by our software. All these control/status lines may be successfully applied in our projects. Moreover, we can expand the number of input/output lines; we will see that shortly.

Interfacing external circuitry to the serial port

Often we directly operate using standard RS-232 signals when designing embedded projects with communications. However, it is more reasonable to deal with TTL-compatible signals; which most of modern electronic devices understands. This simplifies the design of the hardware part of our projects, so we must invent some simple interface to convert RS-232 signals into TTL-compatible ones. There are special chips that greatly simplify this task. The most popular among them is MAX-IM's MAX232 which is "de-facto" acknowledged as an industry standard; we will use this chip in most of our circuits where RS-232 is involved.

Let's see how this converter functions. According to the RS-232 standard, signals on external pins of the serial port connector range in potential from -25 to +25V. With this configuration a logical "0" corresponds to a positive voltage above +3V, while a logical "1" corresponds to signals with a negative voltage below -3V. In practice the RS-232 interface uses a range of -12V through +12V. We can connect the MAX232, according to the scheme shown in Fig.3.12.

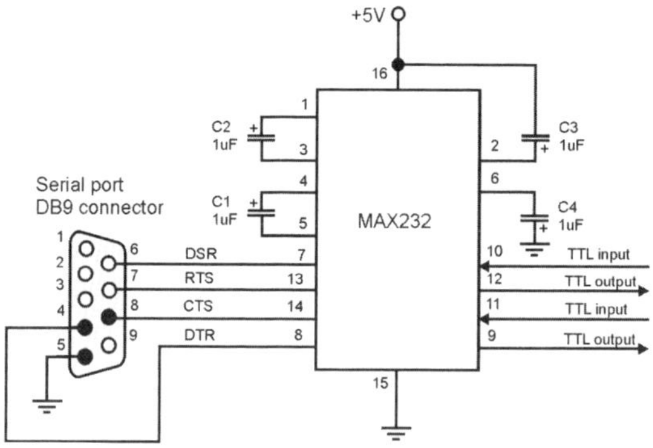

Fig.3.12

For proper operations this circuit requires four capacitors which are needed for an internal voltage conversion from +5V to provide ± 12V supplies. As you can see, signals on pins 9–12 of the MAX232 chip are TTL-compatible, so it is possible to attach external TTL-compatible circuitry there. The MAX232 requires a single power supply of +5V. Beside this chip there are many other modern chips which provide the same function, so you can alternatively apply them in your designs with their appropriate connections.

As you can see, four lines (RTS, DTR, CTS, and DSR) are available for a custom design. To use these signals in embedded projects we need some low-level programming. This may be the best and simplest solution, especially when using the `inb()` and `outb()` functions. Now we need to dig into the serial port hardware in detail. There are many resources describing serial port hardware; you can easily find them when surfing the Internet. I will provide you a brief introduction of the hardware – it will suffice to start off the projects we will discuss.

As we deal with the serial port hardware, we need access to the hardware registers of the UART. Since we apply four control/status lines we must use the two registers MSR and MCR. Both registers were described earlier in this chapter. Recall, that RTS and DTR lines will be manipulated via MCR, while CTS and DSR may be read through MSR. If a base address of a serial port `/dev/ttyS0` is 0x3F8, the MCR's address will be "base + 4" which is 0x3FC and MSR's address ("base + 6") will be 0x3FE.

The feature of all control/status lines of the UART is that when they are low the corresponding bit in MCR/MSR register is set. Again, a 'high' voltage level on some of these lines corresponds to zero bits in the appropriate positions. From this point of view it would be worthwhile and logical to

design our circuits so that low level signals on control/status lines correspond to the appropriate bits set to 0. Vice versa, high level signals correspond to the appropriate bits set to 1.
That is why I slightly modified the above circuit now as shown below in Fig.3.13.

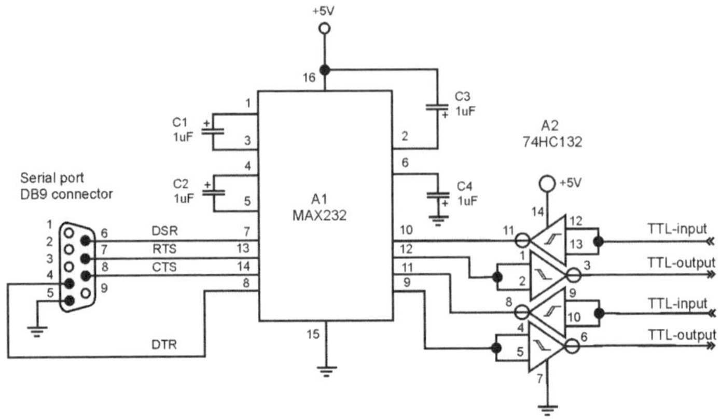

Fig.3.13

Here an additional buffer has been introduced. I used the 74HC132 inverter which, besides an additional inversion of the signals buffers the MAX232 outputs from external circuitry. My assembled circuit is shown below (Fig.3.14):

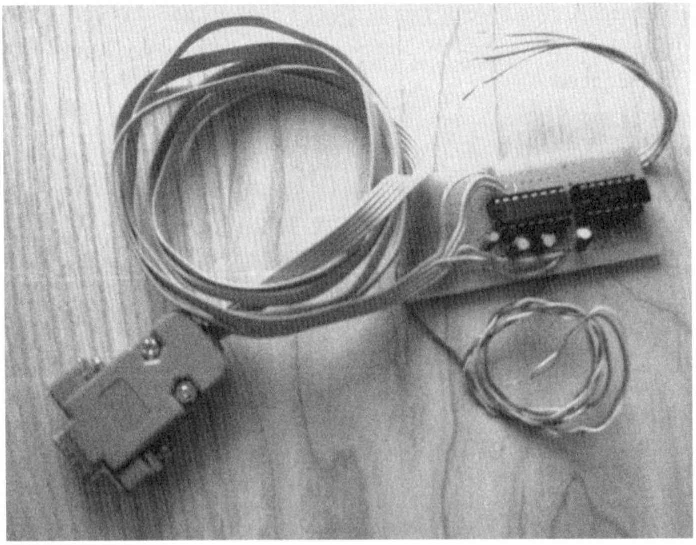

Fig.3.14

Note that the length of an RS-232 cable from the DB-9 connector and our circuit may extend to 12 m. Let's see how to program the control lines. You already know from the previous chapter how to deal with hardware registers of a parallel port. The same is true for serial port hardware. We will mainly use four functions: `ioperm()`, `iopl()`, `inb()` and `outb()`. Recall, that the first two grant access to the hardware from a user application, while `inb()` and `outb()` provide reading/writing from/to the hardware ports. In order to become more familiar with the serial port hardware let's consider the details of two simple demo projects.
The following project shows how to program the RTS control line in gcc. Type the following C source code (Listing 3.3):

Listing 3.3

```
#include <unistd.h>
#include <sys/io.h>

#define MCR 0x3FC  // Modem Control Register = base + 4

int main(void)
{
  char bRead = 0x2;
  iopl(3);
  while (1)
  {
    outb(bRead, MCR);
    sleep(5);
    bRead = ~bRead;
  }
  return 0;
 }
```

This application toggles the RTS line (bit 2 of MCR) every 5 seconds. As the program requires root privileges when starting, you must use the `sudo` command to launch a program file where the above code resides. In order to observe the result, touch one probe of a multimeter (set as a volt meter) to pin 7 (RTS line) of the DB-9 connector and the second probe to pin 5. After the program has started you will see the voltage swing in the range of about 0V to +5V on pin 3 of the 74HC132 inverter. This is a means by which we can also test the DTR output – in this case we must change the initial value of the `bRead` variable (see Listing 3.3) from 0x2 to 0x1:

```
char bRead = 0x1;
```

Save the source code in the `RTS_Test.c` file and compile the source with the following command:

```
$ gcc -o RTS_Test RTS_Test.c
```

To launch the application you should enter:

```
$ sudo ./RTS_Test
```

The following project illustrates how to read the CTS input line. Before testing CTS we need to short RTS (pin 3 of inverter A2) and CTS (pin 9–10 of A2) lines as shown in Fig.3.15:

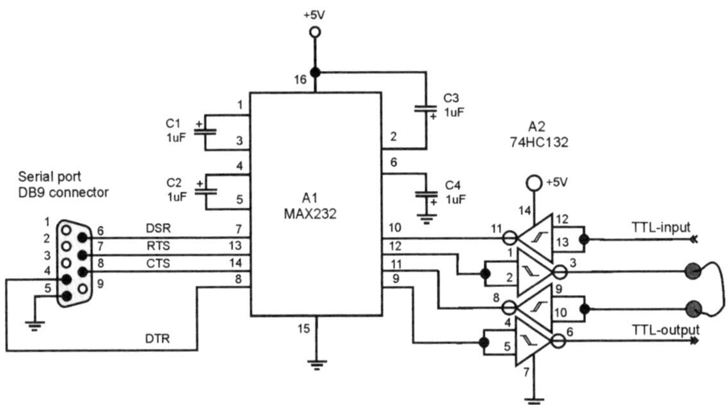

Fig.3.15

The control application will toggle the RTS line and read the appropriate state of CTS. As the RTS and CTS lines are shorted together , the data output on the RTS line immediately appears on CTS . The C source code of a program able to read CTS is shown below in Listing 3.4.

Listing 3.4

```
#include <unistd.h>
#include <stdio.h>
#include <stdlib.h>
#include <pthread.h>
#include <sys/io.h>
```

```
#define MSR 0x3FE    // Modem Status Register = base + 6, bit 4 =
CTS
                     // for /dev/ttyS0

char bRead;
void *counter(void);

main()
{
 pthread_t thread1;
 int  tHandle;

 iopl(3);
 // Creating independent thread

 tHandle = pthread_create(&thread1, NULL, counter, NULL);

 // Wait till threads are complete before main continues. Unless
we
 // wait we run the risk of executing an exit which will terminate
 // the process and all threads before the threads have completed.

 pthread_join( thread1, NULL);
 exit(0);
}

void *counter(void)
{
 char tmp = 0x0;
 while (1)
 {
  do {
      bRead = inb(MSR) & 0x10;
     } while (tmp == bRead);
  tmp = bRead;
  printf("CTS = 0x%X\n", bRead);
 }
}
```

This program uses a separate thread to check CTS. Initially the program creates a new thread by invoking the `pthread_create` function. This function fills the structure `thread1` and assigns the `counter` function responsible for reading the CTS state. The `pthread_join` function allows terminating of the thread before the `main` program has ended.
The `counter` function continuously reads bit 4 of MSR into the `bRead` variable. If the state of this bit has changed the state of CTS is output to the PC screen.
Save the source code in the `Read_CTS.c`. As we use threads here, the command line for the compilation will look as follows:

```
$ gcc -o Read_CTS -lpthread Read_CTS.c
```

In order tp test CTS is functioning we can perform the following steps:
1. Short pins 3, 9-10 of the 74HC132 inverter (see Fig.3.14);
2. Open a terminal window and start the program `RTS_Test` to toggle the RTS line by typing the following command:

   ```
   $ sudo ./RTS_Test
   ```

3. open another terminal window and launch the `Read_CTS` program:

   ```
   $ sudo ./Read_CTS
   ```

To check the DSR line we slightly modify some statements in a `while()` loop of the `counter` function in the source code just discussed (alterations shown in bold in the following listing):

```
while (1)
 {
  do {
      bRead = inb(MSR) & 0x20;
     } while (tmp == bRead);
  tmp = bRead;
  printf("DSR = 0x%X\n", bRead);
 }
```

Beside that we must short pins 3, 12 and 13 of the A2 inverter (see Fig.3.15).
With these four lines RTS, DTR, CTS and DSR we can build a variety of control and measurement systems. By using one control (output) line (RTS or DTR) and one status (input) line (CTS or DSR) we can build simple conditioning systems say to keep some physical parameter (temperature, humidity, pressure, etc.) within a predefined range.

Simple thermostat system

This project describes a simple thermostat system using the LM35DZ temperature sensor which controls the heater of an oven. The schematic circuit for the given system is shown in Fig.3.16:

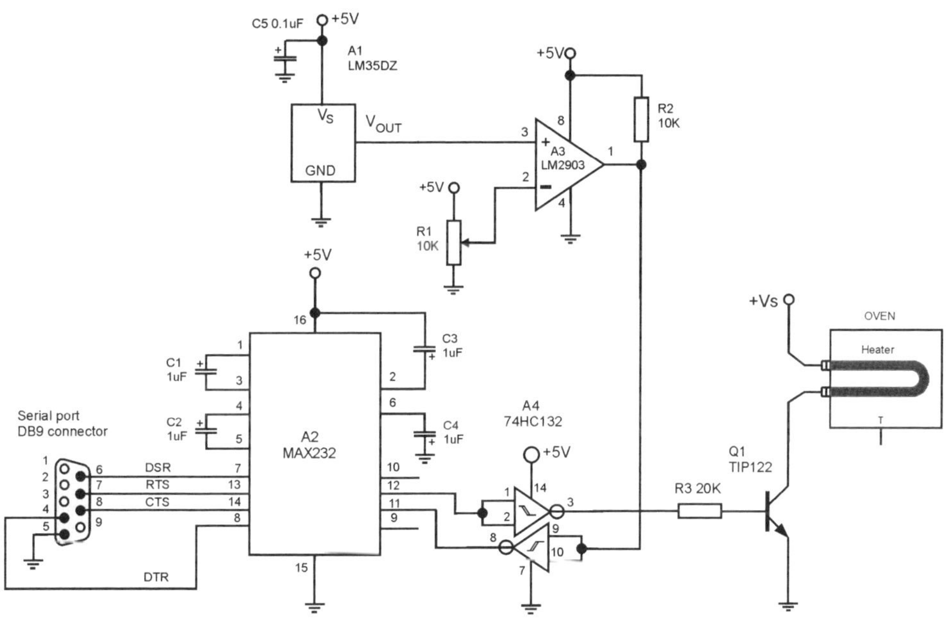

Fig.3.16

A similar circuit was discussed when we were building a similar project based on use of the parallel port. But this circuit uses the serial port lines RTS and CTS instead of parallel port signals. The RTS signal drives the oven by switching on/off the heater, while CTS checks the temperature threshold point. Below is provided a C source code for this project (Listing 3.5).

Listing 3.5

```
#include <stdio.h>
#include <sys/io.h>
#include <unistd.h>

int main(void)
{
 int MCR = 0x3FC; // Modem Control Register = base + 4
 int MSR = 0x3FE; // Modem Status Register = base + 6

 int CTS;

 iopl(3);
 while(1)
   {
     CTS = inb(MSR) & 0x10;
     if (CTS != 0)      // if temperature exceeds threshold
      outb(0x0, MCR);   // the heater should be off, else it will be
                        // turned on
     else
      outb(0x2, MCR);
     sleep(20);
   }
 return 0;
}
```

Let's discuss this source code. The control algorithm is implemented in the endless `while()` loop. Each iteration begins by reading the Modem Status Port (MSR) and saving the byte read in the `CTS` variable. If `CTS` is non-zero that means the temperature has exceeded the threshold, so the heater must be switched off by executing the statement

```
outb(0x0, MCR);
```

The zero value of `CTS` indicates the temperature is below the threshold, so the heater must now be be switched on:

```
outb(0x2, MCR);
```

The last statement within a `while()` loop is the `sleep(20)` function which performs a 20 second delay before repeating the cycle. This delay can be selected randomly by the developer to suit.
The electronic components chosen for this circuit may be easily substituted with similar parts. For example, instead of using the LM2903 chip you could use a single-supply comparator with a TTL output. If you want to reach a higher precision, consider use of a comparator with a low offset voltage – this minimizes the error introduced by the comparator when configuring a threshold voltage.
It is worthwhile to use a high precision potentiometer for R1 – which also affects the accuracy of this circuit. The best result will be achieved in using a specialized single-chip voltage reference instead of R1. When designing this project you should also evaluate the requirements of your real load driven by Q1 and select an appropriate power transistor. Additionally, common precautions must be taken into account when wiring devices to the (vulnerable!) serial port.
Finally, if you are going to use some expansion board with the serial interface, don't forget to replace the addresses 0x3Fx in the source code with appropriate values.
The circuit shown in Fig.3.15 drives a DC load. If we want to on/off control an AC load we must change our circuit. One of the possible solutions is shown in Fig.3.17:

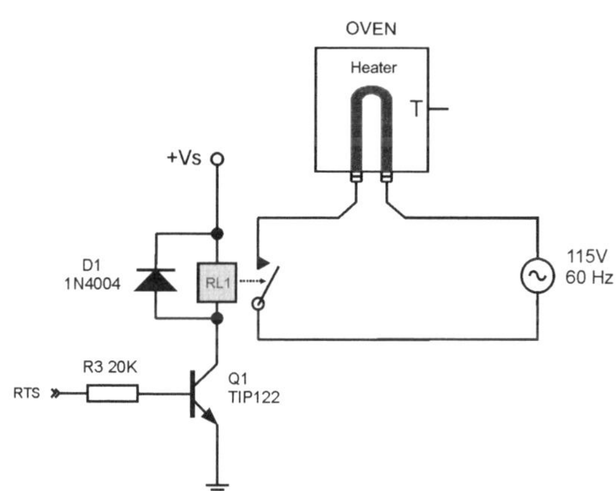

Fig.3.17

Here the Q1 transistor drives the reed relay RL1. Each time the RTS line becomes high Q1 is turned on and RL1 is energized. As a result, the relay contact closes an AC circuit thus connecting power for the heater. When the RTS signal goes low, the Q1 transistor cuts off, thus breaking power to the RL1 relay which, in turn, stops feeding the heater. The 1N4004 diode protects Q1 from high back-emf voltages from the relay coil during cut off of the transistor. With this circuit you can apply any low-power reed relay. An alternative is to use an optocoupler circuit with an optotriac which, in turn, will drive a power triac. One of the possible practical approaches is shown in Fig.3.18:

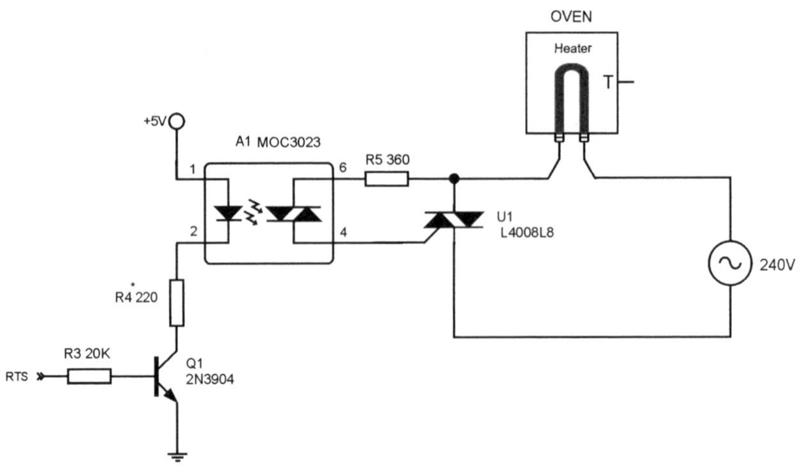

Fig.3.18

The reed relay RL1 has been replaced with an optotriac chip, type MOC3023. This increases reliability (no more mechanical parts moving and wearing) and offers attenuation of electromagnetic noise generated when using a relay. Any of the devices MOC3020 – MOC3023 may be applied here. Another tip; If you want to decrease electromagnetic noise you can include a snubber RC-subcircuit. I encourage you to consult the MOC3020-23 datasheet for other possible modifications. The resistance of R4 must be calculated to provide a suitable on-current through the LED of the optotriac with a range from a maximal value (15mA for MOC3021, 10mA for MOC3022 and 5mA for MOC3023) to an absolute maximum value that is about 60mA.

Naturally, you may apply the above circuits to handle any other similar resistive devices (coolers, bulbs, etc.). For each particular system you should choose the appropriate power triac (U1 in Fig.3.18) to drive your load. If you wish to drive an inductive load (motor, valve, etc.) this is a special case which must be designed for; the discussion of this topic is beyond the scope of this book, so I encourage you to consult additional literature in power electronics.

The projects previously discussed involve the RTS and CTS lines, although we can also use DTR (output) and DSR (input) as well. However, many embedded projects may require more signal lines than the standard serial interface can provide. In this case we can expand the number of input or output lines to gain the necessary functionality. Let's look at the projects below.

Expanding input lines of the serial port

Many times you may need more inputs to capture signals from external circuits than the serial port can provide. The simplest way to expand the number of inputs is to use a multiplexer which can redirect multiple input signals to a single output. The following embedded project demonstrates this approach. The hardware part of the project is represented in the following schematic circuit (Fig.3.19):

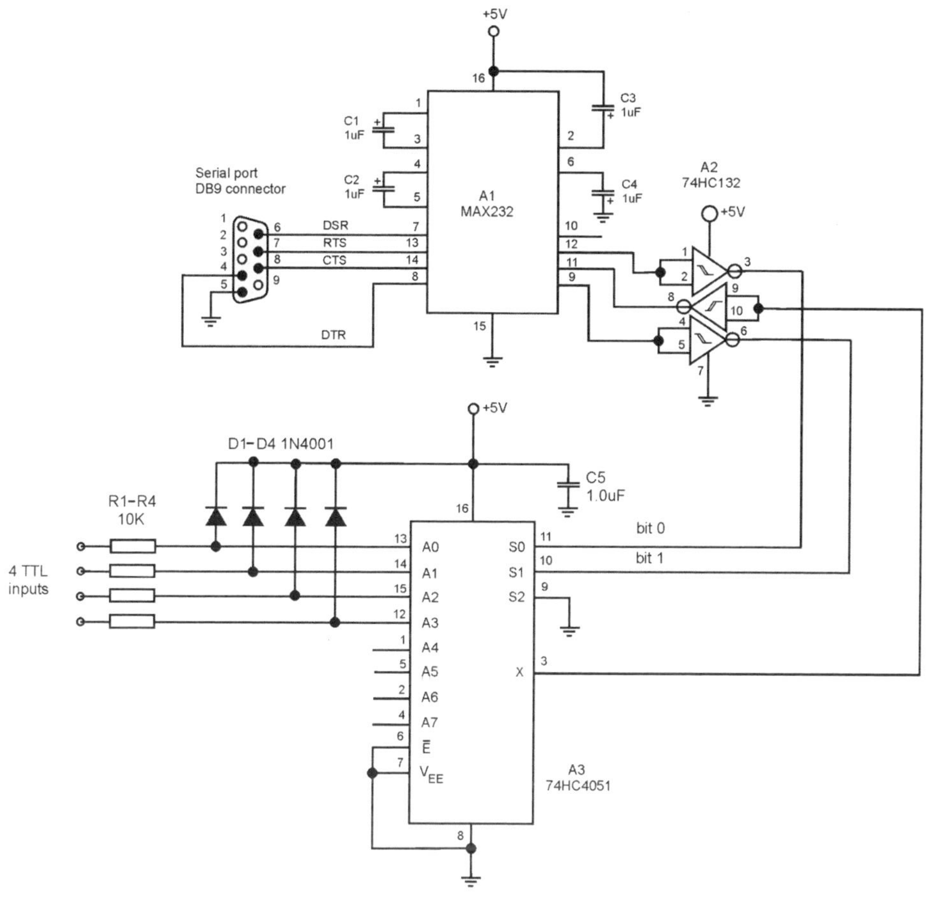

Fig.3.19

This circuit gains 4 additional inputs. As you can see a multiplexer, type74HC4051, was used. With this circuit it is possible to read any one of the A0-A3 digital inputs at a time. The TTL-compatible signal of the input selected goes through the single X output of the 74HC4051. The X output, in turn, is attached to the CTS input line. Thus, the control program will 'know' the state of the X output by reading the state of the CTS line. The program selects an input channel for reading by setting the 2-bit digital code on the S0–S1 inputs of the 74HC4051. For example, setting a code 0x0 on S0–S1 allows reading of A0; code 0x1 allows reading of A1, etc. The S0–S1 inputs are controlled through the RTS and CTS lines of the serial port. Some words about design. The resistor–diode pairs (RxDx, x = 1…4) protect the multiplexer's inputs from higher voltages. Also it is necessary to provide supply noise bypass on the power supply. An adequate bypass is a 1µF capacitor connected as close as possible to the power pins of the multiplexer. As an alternative to the 74HC4051, you can use any 4051 multiplexer which will accept TTL signals. Let's now talk about the software. The control application has the following source code (Listing 3.6):

Listing 3.6

```
#include <stdio.h>
#include <unistd.h>
#include <sys/io.h>
#include <stdlib.h>

int MCR = 0x3FC; // Modem ControlRegister
int MSR = 0x3FE; // Modem Status Register
int DTR;         // DTR (bit 0 of MMCR) = 0x1 when set
int RTS;         // RTS (bit 1 of MCR)  = 0x2 when set
int CTS = 0x10;  // CTS corresponds to bit 4 of MSR being set

int result = 0;

void readBit(int ch);

int main(void)
{
 int channel;
 int tmp = 0;

 iopl(3);
 while (1)
 {
  // reading channel A3 of the 4051 multiplexer
  RTS = 0x2;          // RTS = 1
  DTR = 0x1;          // DTR = 1
  channel = RTS + DTS;
  readBit(channel);

  // reading channel A2
  RTS = 0x0;          // RTS = 0
  DTR = 0x1;          // DTR = 1
  channel = RTS + DTS;
  readBit(channel);
```

```
    // reading channel A1
    RTS = 0x2;          // RTS = 1
    DTR = 0x0;          // DTR = 0
    channel = RTS + DTS;
    readBit(channel);

    // reading channel A0
    RTS = 0x0;        // RTS = 0
    DTR = 0x0;        // DTR = 0
    channel = RTS + DTS;
    readBit(channel);

    printf("Input states A3---A0: 0x%X\n", result);
    sleep(2);
    result = 0x0;
  }
  return 0;
}

void readBit(int ch)
{
    outb(ch, MCR);
    tmp = inb(MSR) & CTS; // select CTS (bit 4 of MSR)
    tmp = tmp >> 4;
    tmp  =  tmp << ch;
    result = result | tmp;
}
```

First of all let's see what variables are used in this program. The `channel` variable holds the actual number of the channel that is being read. As we use two signal lines to select the channel, the maximum number of inputs processed will be $2^2 = 4$. This means that A0-A3 channels of the 4051 may be read. The `tmp` variable is assigned a value indicating the state of the input just read. The following statement from within the `readBit` function does this:

```
tmp = inb(MSR) & CTS;
```

The next two statements put `CTS` into the appropriate position:

```
tmp = tmp >> 4;
tmp  =  tmp << ch;
```

As `tmp` has been calculated, we can place an actual CTS bit into the `result` variable. The following statement saves this bit in the `result` variable:

```
result = result | tmp;
```

After all external inputs have been read, the program outputs a result onto the screen with the `printf()` function. The delay between passes is chosen here to be equal to 2 second (`sleep (2)` function). With this circuit we can read a total of 5 external input lines (A0–A3 inputs of the multiplexer and the DSR input). That may be enough for testing simple digital circuits or reading digital codes from sensors.

Below you can see another solution which also expands the number of serial port inputs. The following project allows adding one additional input to the embedded system. The schematic circuit of this project is shown in Fig.3.20:

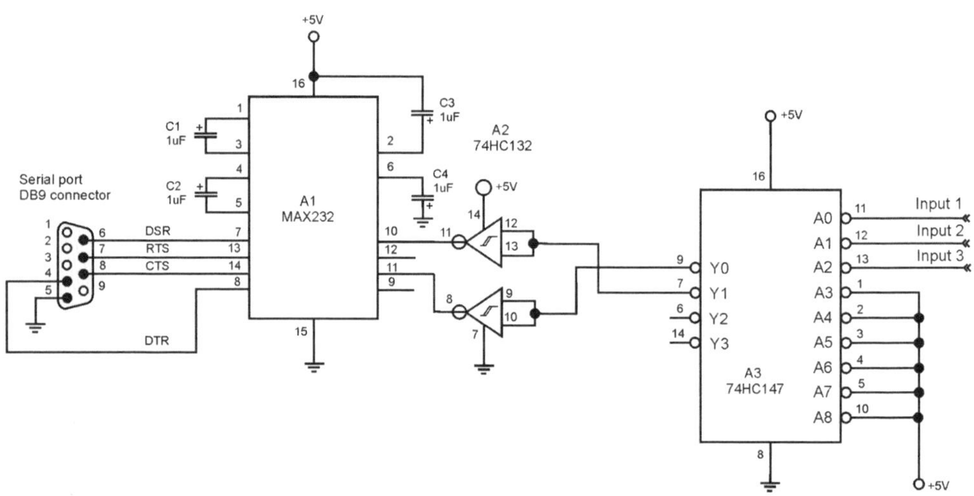

Fig.3.20

Here a 10-to-4 line priority encoder, type 74HC147, is employed. This circuit allows reading up to 3 TTL inputs (Input 1 – Input 3 in Fig.3.20). The 74HC147 priority encoder accepts data from the 3 active low inputs (A0 to A2) and provides a binary representation of the input read on the two active low outputs (Y0-Y1). A priority is assigned to each input so that when two or more inputs are simultaneously active, the input with the highest priority is represented on the output, with input line A2 having the highest priority. The encoder provides the 10-line to 4-line priority encoding function by use of the implied decimal "zero". The "zero" is encoded when all nine data inputs are high, forcing all four outputs high.

The outputs Y0 – Y1 are read through the CTS and DSR RS232 input lines. Recall that the CTS signal is tied to bit 4 of MSR and DSR corresponds to bit 5 of the same register. Thus, to get the state of one of the A0-A2 lines we need to read the MSR hardware register and examine bits 4-5.

Below you can see C source code for the control application (Listing 3.7).

Listing 3.7

```
#include <stdio.h>
#include <unistd.h>
#include <sys/io.h>
#include <stdlib.h>


int main(void)
{
 int MSR = 0x3FE; // Modem Status Register
 int result = 0;

 iopl(3);
 while (1)
 {
  result = inb(MSR); // reading MSR and saving byte read in the
                     // tmp variable
  result = result >> 4;
  result = result & 0x3;
  if (result != 0x0) // check if some of TTL inputs is active
   {
    printf("The highest priority active input is A%X\n", result);
   }
  sleep(2);
  result = 0x0;
 }
 return 0;
}
```

The source code is extremely simple. The program periodically (every 2 s) reads the A0–A2 inputs in a `while()` loop. Each iteration begins in reading a byte from the inputs and saving it in the `result` variable. The following statement does this:

```
result = inb(MSR);
```

As we want the data byte saved in `result` to correspond to the number of the input (1, 2, 3) we need to convert the byte in `result` into an appropriate format by executing the following section of code:

```
result = result >> 4;
result = result & 0x3;
```

After executing these statements the `result` variable will contain one of the possible values in the range of 0 – 3. We do not need '0', so we check for that condition using an `if()` statement. If `result` holds 1 through 3, the program displays the number of the highest priority active input on the screen. That's all about this program.

Expanding output lines with decoder

Another problem that arises when designing embedded projects is a lack of output lines. As you recall, the standard serial port has two control lines, RTS and DTR, which may be used for handling external circuits. The following project illustrates how to expand these 2 lines up to 4. The schematic circuit of this project is shown in Fig.3.21:

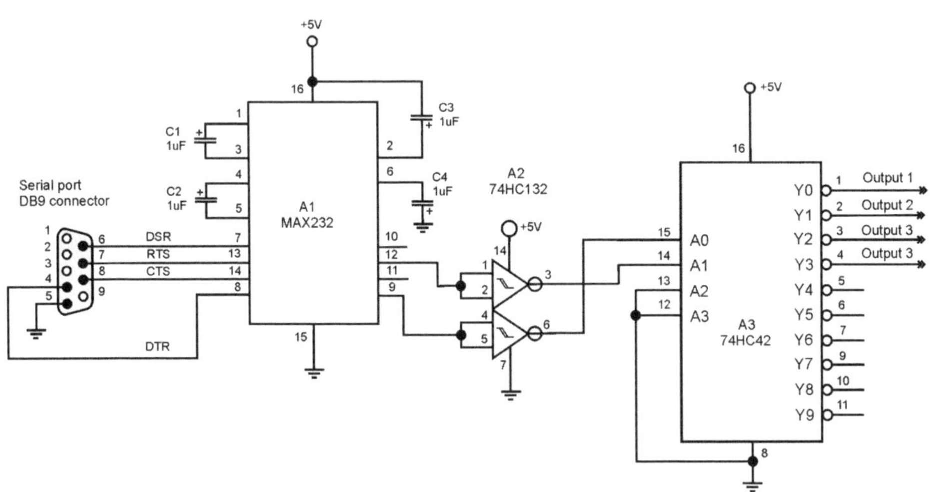

Fig. 3.21

For an expander chip we use the BCD-to-decimal decoder (1-of-10) type 74HC42. The 74HC42 decoder accepts four active high BCD inputs and provides 10 mutually exclusive active low outputs. The logic design of this chip ensures that all outputs are high when binary codes greater than nine are applied to the inputs. In our scheme we use two (A0 and A1) inputs which are driven by the DTR and RTS signals, respectively. Thus we are capable of gaining $2^2 = 4$ outputs that range from Y0 to Y3. Let's see the source code of the demo control application (Listing 3.8).

Listing 3.8

```
#include <stdio.h>
#include <unistd.h>
#include <sys/io.h>
#include <stdlib.h>

int main(void)
{
int MCR = 0x3FC; // Modem ControlRegister
int chOutput = 0;

iopl(3);
while (1)
{
  outb(chOutput, MCR);
  sleep(10);
  chOutput++;
  if (chOutput > 3)
    break;
}
 return 0;
}
```

This program sequentially toggles outputs Y0-Y3 with a 10 second delay between switching.. Setting A0-A1 lines is implemented in an `while(1)` loop. The following statement

```
outb (chOutput, MCR);
```

sequentially writes the values in the range 0 to 3 into the MCR register. This code effects the A0-A1 inputs of the decoder and causes switching on an appropriate output of Y0-Y3. With this circuit we gain 2 inputs (CTS and DSR) and 4 outputs for our embedded system.

Here we discussed some means of expanding inputs/outputs of a serial port. Naturally, you can easily develop I/O expanders using other approaches, for example, shift registers, etc.
So far we have dealt with digital electronics; we developed projects where digital signals were involved. Let's see how to operate with systems where we need to process analog signals.

Processing input analog signals

This next project illustrates how to measure an analog signal represented as a time varying continuous voltage. Usually we deal with analog signals when measuring some physical quantities (temperature, light, pressure, etc.) using analog sensors. The hardware part of this project includes a 12-bit successive approximation (SAR) analog-to-digital converter, type MCP3201, wired to the PC serial port via an SPI-compatible interface (Fig.3.22):

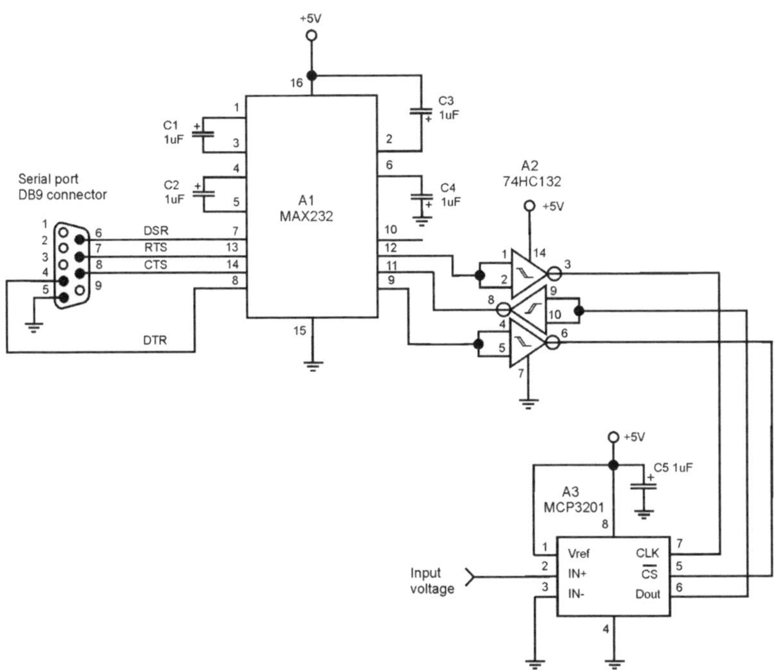

Fig.3.22

The SPI interface involves three signal lines, CS, CLK and Dout. Analog-to-digital conversion (ADC) and data transfer are controlled via those lines by the program running on the PC. The analog input signal in the range of 0 – +5V enters the input (pin 2) of the MCP3201 chip. The CS control line frames the conversion process and must be low while the conversion is in progress. The program controls this line via the DTR bit of the MCR (pin 4 of the DB–9 connector). The output bit stream appears on the Dout pin connected to the CTS line (pin 8 of the DB–9 connector).

The software reads each bit of this bit stream (MSB first) and saves it into an integer variable. After all data bits have been captured the application takes the lowest order 12 bits which represent the original analog signal. The clock pulses synchronizing the data transfer are provided on the CLK pin connected to RTS. The ADC yields each separate bit on Dout on the falling edge of CLK; this bit should be latched by the control application on a rising edge of the clock pulse. To complete the conversion process the control program must then raise the CS line.
A full conversion process is represented by the timing diagram shown in Fig.3.23:

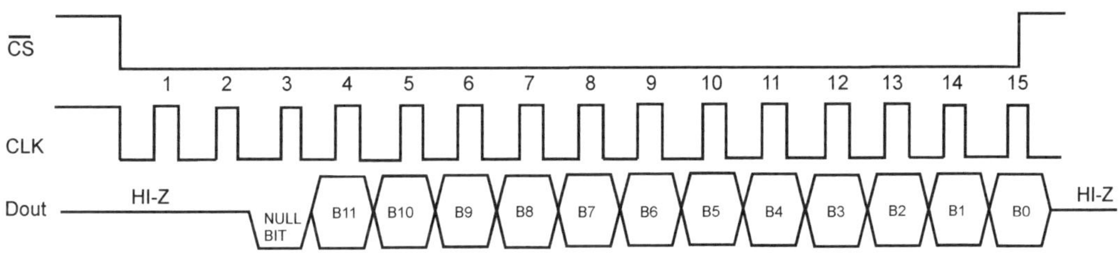

Fig.3.23

Note, that the MSB bit first appears on the Dout line after the 3rd CLK pulse has gone low. As you can see, a full conversion cycle takes 15 CLK pulses, while a data transfer requires 12 cycles.
The circuit should be powered from a +5V 'noiseless' power supply. The bypass capacitor, 10 uF, should be placed as close as possible to the power pin of the MCP3201 chip. If desired, the MCP3201 may be replaced with a similar alternative successive approximation ADC having an SPI-compatible interface. For instance, you may use a LTC1286 or LTC1297 device (Linear Tech.), etc. In this case you need to reconsider all circuit connections and modify the source code according to the timing diagram of your device selected.
Let's discuss the software. The source code of our control application is presented below (Listing 3.9):

Listing 3.9

```
#include <stdio.h>
#include <unistd.h>
#include <stdlib.h>
#include <sys/io.h>

#define MCR 0x3FC  // Modem Control Register address (base + 4)
#define MSR 0x3FE  // Modem Status Register address  (base + 6)
```

```
int main(void)
{
 int i1;
 int binData = 0;
 int mask = 0x4000;
 int dataCTS = 0;
 double total = 0;

 iopl(3);               // CS   is a Bit 0 of MCR,
                        // CLK  is a Bit 1 of MCR,
                        // Dout is a Bit 4 of MSR

 outb(0x1, MCR);        // CS goes high
 outb(0x3, MCR);        // CS stands high, CLK goes high
 outb(0x2, dataPort); // CS goes low thus allowing the conversion

for (i1 = 0; i1 < 15; i1++)
 {
  outb(0x0, MCR);                               // CLK goes low
  dataCTS = inb(MSR) & 0x10;                    // CTS bit is selected
  if (dataCTS != 0)
    binData = binData | mask;
  outb(0x2, MCR);                               // CLK goes high
  mask = mask >> 1;
 }
 outb(0x3, MCR);                                // CS goes high
 binData = binData & 0xFFF;
 total = 5.00 / 4096 * binData;
 printf("Input voltage is: %5.3f V\n", total);
 return 0;
}
```

First of all, take a look at the variables used. The `dataCTS` variable temporarily holds the CTS bit just read via the Dout line. The `binData` variable collects the binary code bit by bit. After the last iteration this variable will hold a binary code of the actual analog signal. Each bit falls into the proper position in `binData` by executing the following statement:

```
binData = binData | mask;
```

Here the `mask` variable helps to place the bit being read into a proper position within `binData`. After executing the above statement, the Dout bits will be arranged in the right places in the `binData` variable.
The `total` variable of a double type holds the final value of the input analog voltage; the `printf()` function outputs this value after the following calculation has been processed:

```
total = 5.00 / 4096 * binData;
```

Here the constant of 5.00 denotes the reference voltage (in volts) applied to pin 1 of the MCP3201 converter. Usually, this reference voltage varies around 5.00V, so you must type an exact value in the above statement. A higher accuracy of analog-to-digital conversion is achieved by using a precision voltage reference; in that case the accuracy of the voltage reference determines the overall accuracy of a given conditioning circuit.
The value of the reference voltage divided by 4096 (2^{12}) gives a size (in volts) of the LSB of the converter. In our case this value will be 5.00 / 4096 = 0.00122V (1.22mV). The program multiplies the binary code held in `binData` by the LSB and stores the result in the `total` variable. That will be an analog voltage on the input of the MCP3201 converter.
The control application implements the timing diagram of the conversion within the `for()` loop. Since the full conversion process requires 15 iterations, the program code sequentially reads 15 bits, from which we are interested only in a first 12 low order bits. For this reason we throw away the higher 3 bits by executing the statement

```
binData = binData & 0xFFF;
```

Save our source code in the `ADC.c` file and compile it by entering the following command:

```
$ gcc -o ADC ADC.c
```

On a successful compilation we can launch the application using the following command:

```
$ sudo ./ADC
```

The project just discussed may be used as a template for designing various embedded system where analog signals are estimated to control some various equipment. The next project illustrates how to design and build a simple pressure measurement system.

Pressure measurement with the MPX4115 sensor

As the sensitive element for our pressure measurement system we will apply the MPX4115A pressure sensor. It was designed to sense absolute air pressure in altimeter or barometer (BAP) applications. This BAP sensor integrates an on-chip, bipolar op-amp circuitry and thin film resistor networks to provide a high level analog output signal and temperature compensation.
The MPX4115A allows pressure measurements from 15 to 115 kPa (2.2 to 16.7 psi) that corresponds to the output voltage 0.2 through 4.8V. As you can see, the transfer function shown in Fig.3.24 has a high linearity.

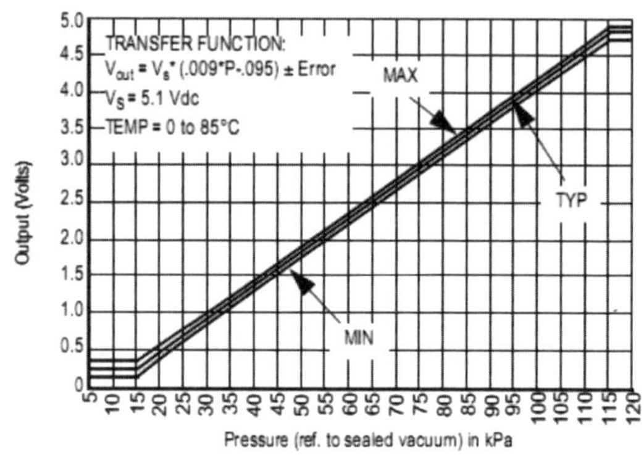

Fig.3.24

The MPX4115A sensor provides the output voltage V_{OUT} that may be calculated as:

$$V_{OUT} = V_S \times (P \times 0.009 - 0.095) \pm (\text{Pressure Error} \times \text{Temp.Factor} \times 0.009 \times V_S),$$

where $V_S = 5.1V \pm 0.25V$.
When the measurements are taken in a temperature range of 0°C — 85°C, the Temp.Factor is unity.
Pressure measurement yields a pressure Error equal to ±3.45kPa in the range of 20 to 250 kPa.
To simplify calculations lets ignore the possible but minimal errors in a given project, thus the above formula may be rewritten as follows:

$$V_{OUT} = V_S \times (P \times 0.009 - 0.095)$$

From this equation we may found out the pressure P:

$$P = (V_{OUT} / V_S + 0.095) / 0.009$$

These equations can be used by software to calculate the absolute pressure in kPa units. To convert the result from kPa into mmHg units (this is an abbreviation of "millimeters of mercury") we should apply the following formula:

$P_{mmHg} = (P \times 7500.7\ ) / 1000,$

where P_{mmHg} is the pressure in "millimeters of mercury" and P – pressure measured in kPa units. Let's examine the schematic circuit of our system (Fig.3.25):

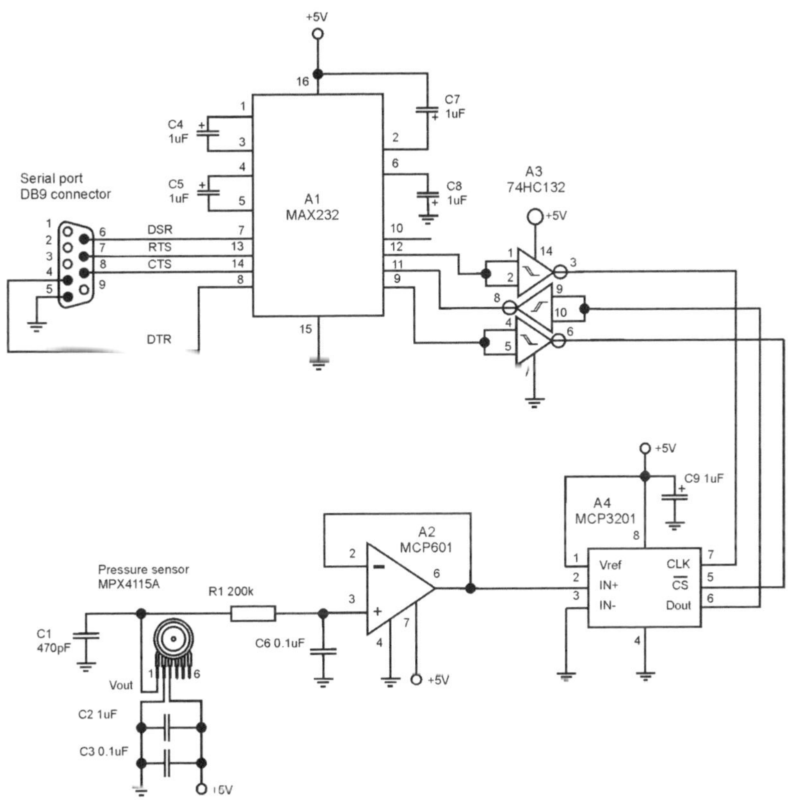

Fig.3.25

The signal chain from the pressure sensor to the MAX232 converter consists of an active low–pass filter which includes operational amplifier A2, type MCP601, and a R1C6 pair. The active low-pass filter is an optional stage; it comes in handy when noise attenuation is needed. The analog signal from the sensor then passes to the converter's input (pin 2 of A4). The conversion process is controlled via an SPI–compatible interface built using the RTS, CTS and DTR signal lines. As usual, the MAX32 chip converts RS-232 signals to TTL-compatible levels.

Let's talk about the physical implementation.
When using analog-to-digital converters, all ground, positive and negative supply pins should be connected to the analog planes. Also, if the converter has an IN- pin associated with the input signal, that pin should be connected as close as possible to the signal ground. The analog and digital power pins should be connected to the analog power plane with proper bypass capacitors placed as close as possible to each pin. The bypass capacitors C1-C3 should be placed as close as possible to the power pin of the sensor. It's desired that the power supply should be noiseless or at least have a minimal noise.
Let's see the software. The C source code for this project is shown in Listing 3.10.

Listing 3.10

```
#include <stdio.h>
#include <unistd.h>
#include <stdlib.h>
#include <sys/io.h>

#define MCR  0x3FC  // Modem Control Register
#define MSR  0x3FE  // Modem Status Register


#define Vref 5.00   // Voltage reference for ADC

int main(void)
{
 int i1;
 int binData = 0;
 int mask = 0x4000;
 int dataCTS = 0;

 double Vout;
 double pressureKPa;
 double pressuremmHg;

 iopl(3);                // CS   is a Bit 0 of MCR,
                         // CLK  is a Bit 1 of MCR,
                         // Dout is a Bit 4 of MSR
```

```
 outb(0x1, MCR);      // CS goes high
 outb(0x3, MCR);      // CS stands high, CLK goes high
 outb(0x2, dataPort); // CS goes low thus allowing the conversion

for (i1 = 0; i1 < 15; i1++)
 {
  outb(0x0, MCR);                          // CLK goes low
  dataCTS = inb(MSR) & 0x10;               // CTS bit is selected
  if (dataCTS != 0)
    binData = binData | mask;
  outb(0x2, MCR);                          // CLK goes high
  mask = mask >> 1;
 }
 outb(0x3, MCR);                           // CS goes high
 binData = binData & 0xFFF;

 Vout = Vref / 4096 * binData;
 pressureKPa = (Vout / Vref + 0.095) / 0.009;
 pressuremmHg = pressureKPa * 7500.7 / 1000.0;
 printf("Pressure = %5.3f KPa (%5.2f mmHg)\n", pressureKPa, pres-
 suremmHg);
 return 0;
}
```

In this program the source code should already be very familiar from the previous project except now the `pressureKPa` variable will hold the value of absolute pressure in KPa units and the `pressuremmHg` variable contains the value calculated in mmHg units.

Expanding the serial interface using I²C

The I²C interface allows expansion of the number of inputs/outputs with which an embedded system can operate. As you already know, external circuits attached to the serial port can operate with four digital lines: two outputs (RTS and DTR) and two inputs (CTS and DSR). Nevertheless, the number of serial port inputs/outputs may be increased by using I²C expanders. The following project shows how to expand two serial port output lines to eight.

Let's examine the hardware first. A schematic circuit of the serial port I²C expander is shown in Fig.3.26:

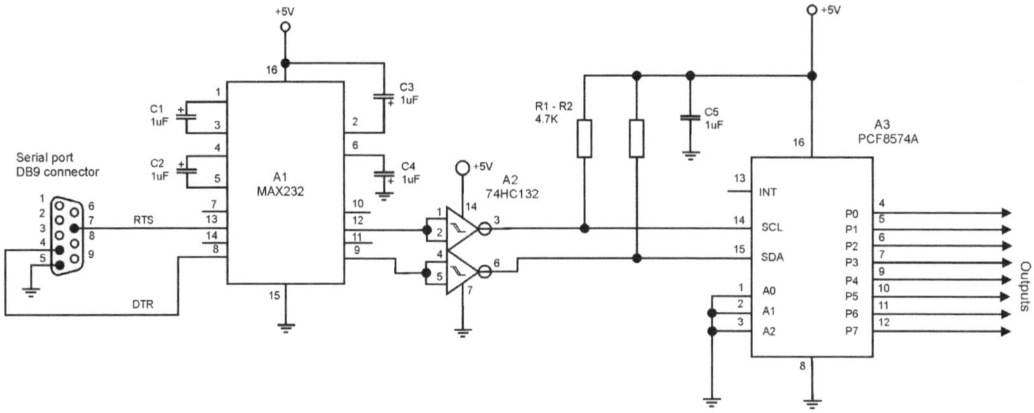

Fig.3.28

This circuit is simple, requiring few components. The MAX232 chip converts RS-232 signal levels from the PC's serial port to TTL-compatible levels needed to operate I²C bus. The I²C interface uses a PCF8574A I²C I/O expander that provides standard SCL and SDA signals controlled by the RTS (ready-to-send) and DTR (data-terminal ready) serial port lines. An 8-bit binary code (data byte) is passed from the serial port of the PC to the I²C I/O expander outputs (P0-P7 pins).

According to the I²C specifications, the data byte passes via the serial data (SDA) signal line, while the synchronization is provided by the serial clock (SCL) line. For a more complete understanding of how the timing diagram of I²C is implemented you may wish to refer to the datasheet of the PCF8574A chip. The A0–A2 address lines define the address of the device on the I²C bus. In a given case those lines are connected to "ground". Remember, that each device on the I²C bus must have a unique address. Anyway, when assigning the address you should consult the device's datasheet. That's all about hardware.

Below you can see the C source code of the control application (Listing 3.11):

Listing 3.11

```
#include <stdio.h>
#include <unistd.h>
#include <stdlib.h>
#include <sys/io.h>

#define MCR  0x3FC  // Modem Control Register for /dev/ttyS0
// RTS <-> SCL
// DTR <-> SDA
```

```
#define SCL 0x2 // high level on the SCL line
#define SDA 0x1 // high level on the SDA line

void I2CStart();
void I2CStop();
void WriteByte(int b1);

int main(void)
{
 int i1;
 int dat;
 int cmd;
 char tmp;

 iopl(3);

 dat = 0x1;
 I2CStart();
 cmd = 0x70;
 WriteByte(cmd);
 WriteByte(dat);
 I2CStop();

 return 0;
}

void I2CStart()
{
 outb(SCL | SDA, MCR);
 outb(SCL, MCR);
}


void I2CStop()
{
 outb(0x0, MCR);        // SCL goes low, SDA goes low
 outb(SCL, MCR);        // SCL goes high
 outb(SCL | SDA, MCR); // SDA goes high while SCL is high
}
```

```
void WriteByte(int b1)
{
 int i1;
 for (i1 = 0; i1 < 8; i1++)
 {
   outb(0x0, MCR);           // SCL goes low
   if ((b1 & 0x80) != 0)  // if MSB of b1 = 1, SDA is assigned 1
     {
      outb(SDA, MCR)
     tmp = SDA;
    }
  else
    {
     outb(0x0, MCR);       // if MSB of bit 1 = 0, SDA is assigned 0
     tmp = 0x0;
    }
  outb(tmp | SCL, MCR);
  b1 = b1 << 1;
 }
 outb(SCL, MCR);
 outb(SCL | SDA, MCR);
}
```

Recall, that the SCL signal line of the I²C interface is controlled by RTS and the SDA line is controlled by DTR from the serial port.
Here two variables, `cmd` and `dat`, are used. The `cmd` variable holds the code of the command that has to be sent to the I²C device in a first byte. The command is a combination of the address of the device and the code of the operation (read/write). In our case, the code of the command will be equal to 0x70. The `dat` variable contains the data byte that appears on P0–P7 outputs after a write cycle is completed. Writing data via the I²C bus for the PCF8574A device takes four steps: issuing the "start sequence", writing the first byte which contains the command, writing the data byte that must be on the P0–P7 outputs and issuing the "stop sequence". These steps will be executed inside the main program that invokes the appropriate functions in the right order. The `I2CStart` function generates the "start sequence". The `WriteByte(cmd)` function passes the first byte (command, `cmd` variable) to the PCF8574A device. Then the `WriteByte(dat)` function passes the data byte which is outputted to the P0–P7 pins. Finally, the `I2CStop` function generates the stop sequence which ends a data transfer via the I²C interface.

The design may be easily improved upon by increasing the number of outputs up to 16 after replacing the PCF8574A chip using a PCF8575C. Also, there are other I²C expanders that may be used in this design, for instance, MCP23017 by Microchip, etc. When you make such replacements you should also change the schematic circuit in Fig.3.28 in an appropriate way and rewrite the source code of the application according to the timing diagram of the selected chip. Instead of a MAX232 line driver, you can use a similar alternative, for instance, the MAX225 or MAX233. These parts don't require any external components, thus simplifying the design.

Using MATLAB in home projects

This section will be dedicated to making use of the popular and powerful development tool called MATLAB®. This application is widely used for building various embedded systems for home and laboratories. As you may know, MATLAB is used for maths and computation, in development of multiple data acquisition systems, for modeling, simulation and prototyping. Other uses include scientific and engineering graphics, application development with a graphical user interface.
MATLAB is an interactive system that allows you to solve many technical computing problems providing C language tools. It is not the intention of this book to provide a detailed description of all capabilities of MATLAB. We will see how to apply MATLAB for embedded projects where a serial interface is used. The advantage of using such a powerful tool as MATLAB in an embedded design is that you need only build your hardware and provide an interface to your device, while other tasks including data processing, maths and visualization of end results will be easily realized inside the MATLAB environment.
To operate with MATLAB you should have some experience in programming in this development environment. For my experiments I used the R2009b Linux version of this product, although older versions suit as well. Detailed instructions on how to configure your MATLAB in Linux may be found on sites relating to Linux and MATLAB.
Let's see how to program data transfer via a serial interface using MATLAB. Consider the general scheme where we have the external device (this may be some measurement device, microcontroller based data acquisition system, etc.) attached to the serial port and controlled by MATLAB. The most common case is when your application sends some command to the external device and receives an answer. Let's go to a practical example. Our embedded system will contain a simple 8051family microcontroller development board (I personally used 8052 board from Rigel Corp.) attached to the serial interface `/dev/ttyS0` (Fig.3.27).

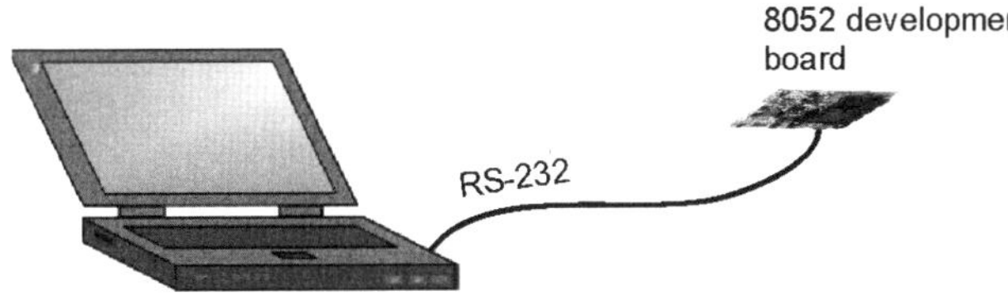

Fig.3.27

To make things clear, suppose our 8052 board runs a simple application based on the C source code (the Keil C compiler is used) shown below in Listing 3.12.

Listing 3.12

```
#include <REG52.H>
#include <stdio.h>
#include <string.h>


void delay(void)
{
 long interval = 500;
 while(interval-- != 0);
}

int main(void)
{
  char buf[128];
  char cmd1[] = "*IDN?";
  char cmd2[] = "*TEST?";
  char data1[] = "8051 BASED SYSTEM";
  char data2[] = "TESTED OK";

  char i;
  int args;

  SCON = 0x50;
  TMOD = 0x20;
  TH1 = 0xF3; //Baud rate 9600 at Fosc = 24.0MHz
  PCON |= 0x80;
  TR1 = 1;
  TI = 1;

  while (1)
  {

   memset(buf, '\0', sizeof(buf));
   args = scanf("%s", buf);
```

```
    if (args == EOF)
     continue;

    i = strncmp (buf, cmd1, 5);
    if (i == 0)
    {
  delay();
  printf("ANSW: %s\n",data1);
     continue;
    }

    i = strncmp(buf, cmd2, 6);
    if (i == 0)
    {
  delay();
  printf("ANSW: %s\n",data2);
  continue;
    }
    else
     {
   delay();
   printf("Wrong command\n");
   continue;
  }
  }
}
```

The 8052 program is extremely simple. The microcontroller receives a data byte from a PC and tries to interpret this data as a command. The 8052 software is developed so that it can recognize two specific commands. When either of them is found, the 8052 program sends an answer back to the PC. For the sake of simplicity no other actions are undertaken besides those mentioned above.

The source code contains two variables `cmd1` and `cmd2` of a char* type that points to the command strings; variables `data1` and `data2` are strings that the 8052 sends back to the PC after one of the `cmd1` or `cmd2` has been found. The `data1` string is sent when the 8052 program recognizes the `cmd1` command, while `data2` is an answer to `cmd2`. Thus, if the "*IDN?" command is sent to the 8052 the answer will be "8051 BASED SYSTEM". Again, after the "*TEST?" string has been received the microcontroller answers "TESTED OK".

The following statements

```
SCON = 0x50;
TMOD = 0x20;
TH1 = 0xF3; //Baud rate 9600 at Fosc = 24.0MHz
PCON |= 0x80;
TR1 = 1;
TI = 1;
```

initialize the serial interface of the 8052. Here, we configure a baud rate of 9600 from the clock frequency of 24.0MHz at which this 8052 module operates.

After initialization has been performed the program starts to implement its main job. Each command received from the PC is compared with the `cmd1` and `cmd2` templates by calling a `strncmp` C library function. After a coincidence match has been found the application sends appropriate data back to the PC. This is how this 8052 software works.

The second part of the software works in the MATLAB environment. You can test how this system works by entering the following sequence in the command window of MATLAB environment (Listing 3.13):

Listing 3.13

```
>> s = serial('/dev/ttyS0');
>> s.timeout = 30;
>> s.timeout

ans =

    30

>> s.BaudRate = 9600;
>> s.BaudRate

ans =

        9600

>> s

   Serial Port Object : Serial-/dev/ttyS0
```

```
   Communication Settings
      Port:                 /dev/ttyS0
      BaudRate:             9600
      Terminator:           'LF'

   Communication State
      Status:               closed
      RecordStatus:         off

   Read/Write State
      TransferStatus:       idle
      BytesAvailable:       0
      ValuesReceived:       0
      ValuesSent:           0

>> fopen(s);
>> s

   Serial Port Object : Serial-/dev/ttyS0

   Communication Settings
      Port:                 /dev/ttyS0
      BaudRate:             9600
      Terminator:           'LF'

   Communication State
      Status:               open
      RecordStatus:         off

   Read/Write State
      TransferStatus:       idle
      BytesAvailable:       0
      ValuesReceived:       0
      ValuesSent:           0

>> fprintf(s, '*IDN?');
>> s.BytesAvailable
```

```
ans =

    32

>> msg = fread(s, 32);
>> fprintf(1, '%s', msg);
Warning: The argument for the %s format specifier must be of type
char (a string).
*IDN?

ANSW: 8051 BASED SYSTEM

>> msg

msg =

    42
    73
    68
    78
    63
    13
    10
    65
    78
    83
    87
    58
    32
    56
    48
    53
    49
    32
    66
    65
    83
    69
```

```
    68
    32
    83
    89
    83
    84
    69
    77
    13
    10
>> fclose(s);
>> delete(s);
>> clear s;
```

This sequence contains some statements which are not mandatory, but they reveal useful information concerning the serial port. The first statement creates a serial port object `s` which represents the given serial interface:

```
>> s = serial('/dev/ttyS0');
```

The `serial` function takes the name of the serial port as a parameter. Additionally, we can also configure property values during an object creation. Each serial port object is associated with one serial port. In our case, the serial port object is associated with a serial port `/dev/ttyS0`. If, for example, `/dev/ttyS0` does not exist, or it is already in use, we will not be able to connect the serial port object to the device.
The following statement

```
>> s.timeout = 30;
```

sets the timeout period for read/write operations. You must configure the `Timeout` option to be the maximum time (in seconds) to wait to complete a read or write operation. If a timeout occurs, the read or write operation aborts. The default value for the timeout was 10 s, but we have changed it to 30. Many times we don't need to change the default value for this parameter; you can leave its value unchanged.
To view the serial port parameters we entered the name of the serial port object without a semicolon:

```
>> s
```

This causes the parameters of the serial interface to be displayed on the screen.
Note that the list of parameters contains the `Status` field which describes the state of a given serial interface. After creating a serial port object with the `serial` function the serial port still remains closed. As we want to operate on a serial port, we need to open the serial port object by calling the `fopen` library function:

```
>> fopen(s);
```

After executing this command the `Status` property has been changed to "open". The following statement writes the command "*IDN?" to the 8052 board via the serial port:

```
>> fprintf(s, '*IDN?');
```

The 8052 device responded by sending the appropriate line back to MATLAB. Before reading the serial port we checked if some data was already available in the input buffer:

```
>> s.BytesAvailable
```

In our case we see that 32 bytes had been received, so we read this data using the `fread` function:

```
>> msg = fread(s, 32);
```

Note, that the `fread` function reads data in a binary format, so the `msg` variable held the string "ANSW: 8051 BASED SYSTEM" as a sequence of binary codes. To view data as a string of characters we can call the `fprintf()` library function:

```
>> fprintf(1, '%s', msg);
```

Another way is to simply enter `msg` without a semicolon at the end:

```
>> msg
```

We need to close the serial port object after operations have completed. The following statements do that:

```
>> fclose(s);
>> delete(s);
>> clear s;
```

The `fclose` function closes the serial port object. Then we remove the `s` variable associated with this object and clear the memory.
This sequence of statements can be saved in a M-file. Next time we want to execute this sequence we simply enter the name of the M-file in the command window.
Let's see, for example, how to switch on/off the AC load in the circuit drawn in Fig.3.7 — 3.8. The task may be accomplished by executing this following sequence:

```
>> s = serial('/dev/ttyS0');
>> fopen(s);
>> fprintf(s, <any character>);
>> fclose(s);
>> delete(s);
>> clear s;
```

Naturally, you can easily improve the above source code in and gain additional functionality. As you hopefully recall, the serial port has four digital lines designated for control operations (RTS, DTR, CTS and DSR). We have considered their use earlier in this chapter, but now lets see how those signals can be processed using MATLAB.
MATLAB deals with control signals by reading/writing the appropriate properties of the serial port object. In the table below you can see the control pins assignments and their properties (Table 3.1):

Table 3.1. Serial Port Control Pin Properties

Property Name	Description
RequestToSend	Specify the state of the RTS signal
DataTerminalReady	Specify the state of the DTR signal
PinStatus	Indicates the state of CD, CTS, DSR and RI lines

We are interested in using two outputs (RTS and DTR) and two inputs (CTS and DSR). As you can see from Table 3.1, we can set/clear RTS and/or DTR by assigning appropriate values to the control pin properties. Setting either of RTS or DTR signal lines is accomplished by assigning "on" to the appropriate property. Otherwise, clearing those signal lines may be performed when assigning them the "off" value. In order to experiment with control pins in MATLAB we need a circuit thus (Fig.3.28):

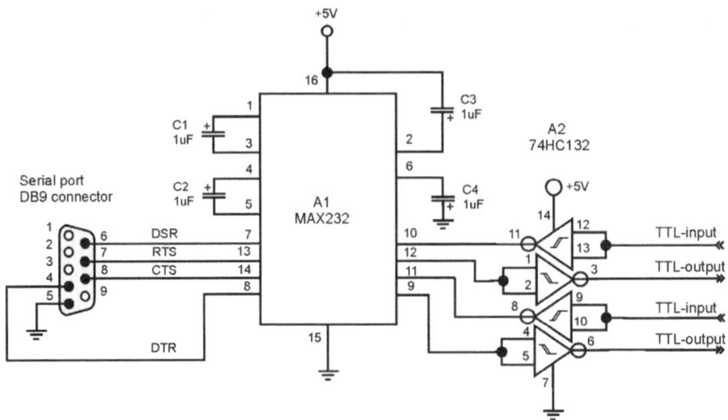

Fig.3.28

This circuit we already used when designing earlier projects in this chapter. With this circuit setting/clearing the RTS or DTR bits in the MCR register causes raising/falling the appropriate TTL output. As to CTS and DSR, the appropriate bit in the MSR register will be set/ cleared when the appropriate line becomes high/low. These conditions simplify programming because a logical state of "1" in the MCR/MSR bit corresponds to the high level on the appropriate line, while a logical "0" results in a low state on the signal line.

If, for instance, the `RequestToSend` property is set to "on" (the RTS bit in MCR is set) the RTS line on pin 3 of 74HC132 inverter goes high. Otherwise, setting the `RequestToSend` property to "off" by clearing RTS bit in MCR causes the RTS signal line to drop. The same is also true for DTR, CTS and DSR signals.

Let's see how we can operate with those control and status pins. After MATLAB has started let's open the `/dev/ttyS0` device and see the state of its input lines. The following sequence does that:

```
>> s = serial('/dev/ttyS0');
>> fopen(s);
>> s.PinStatus

ans =

    CarrierDetect: 'off'
    ClearToSend: 'off'
    DataSetReady: 'off'
    RingIndicator: 'off'
```

As you can see, we have got information about the input lines of a given serial interface.

If, for example, we are interested only in getting the status of CTS (`ClearToSend` property) or DSR (`DataSetReady` property) we may enter the following commands:

```
>> s.PinStatus.ClearToSend

ans =

off

 >> s.PinStatus.DataSetReady

ans =

off
```

These commands are often used in M-files when reading the state of input lines of the serial interface.

As to control lines, RTS and DTR, their status may be determined by typing the commands shown below. The statement

```
>> s.RequestToSend

ans =

off
```

allows us to see the state of RTS ("`off`" in that particular case), while

```
>> s.DataTerminalReady

ans =

on
```

statement gives us the state of DTR ("`on`" for a given example).

One more complicated example of using control lines is presented below. This example exhibits how to periodically trigger your serial interface to accomplish some useful task. Here the RTS line will be switched on/off after some time has passed. The timing in this project is provided by a timer object. After some interval expires, the timer is activated and invokes the callback function which executes some task.

The program code within this function checks the state of the RST line and reverses the voltage level on RTS. The source code of a timer callback function named as an `rts_trigger` is shown below (Listing 3.14):

Listing 3.14

```
function rts_trigger(obj,event)
%UNTITLED2 Summary of this function goes here
%   Detailed explanation goes here

global s;
i1 = strcmp(s.RequestToSend, 'on');
if (i1)
    s.RequestToSend = 'off';
else
    s.RequestToSend = 'on';
end;
end
```

The variable `s` is declared as global, so as to be visible for other applications. Here `s` will present the serial port object, through which the callback function gets control over a RTS line of the serial interface opened by the other application. The following statement compares the value of the `RequestToSend` property of the serial port object `s` with the "on" string and returns the result as a logical value of 1 (true) or 0 (false) in the variable `i1`:

```
i1 = strcmp(s.RequestToSend, 'on');
```

The`if...else` statement that follows reverses the RTS line. This source code should be saved in an M-file `rts_trigger.m` (note: the file name must be the same as one of the callback function). The main program may be typed in the built-in editor of MATLAB and saved in some M-file that you launch when needed. In our case I simply entered the following commands in the command window without saving:

```
>> t = timer('StartDelay', 15, 'Period', 5, 'TasksToExecute', 10,
'ExecutionMode', 'fixedRate');
>> t.StartFcn = 'disp(''TIMER started.'')';
>> t.StopFcn = 'disp(''TIMER stopped.'')';
>> t.TimerFcn = 'rts_trigger';
>> global s;
>> s = serial('/dev/ttyS0');
>> s.RequestToSend
```

```
ans =

off

>> fopen(s);
>> s.RequestToSend = 'on';
>> start(t);
TIMER started.
TIMER stopped.
>> delete(t);
>> clear t;
>> fclose(s);
>> delete(s);
>> clear s;
```

The very first statement above creates a timer object `t` and assigns to it some properties. Note that the timer object `t` when configuring takes the pairs "property-value", where "value" is assigned to "property". The "`StartDelay`" property indicates time which passes before the timer triggers for a first time – we set it to 15 s. The "`Period`" property defines the delay between events; we select 5 seconds. Then the "`TasksToExecute`" follows. it determines the number of times the timer activates – we put this value to 10. The next property we need to define is "`ExecutionMode`". This property determines how a timer object schedules timer events. We define this property as "`fixedRate`".
The next steps we need are to define functions which are invoked during starting, running and stopping of the timer. Two following commands define the function which is invoked while the timer starts off:

```
>> t.StartFcn = 'disp(''TIMER started.'')';
```

There will be a `disp` function which outputs a "TIMER started." string. The other `disp` function provides the output of a "TIMER stopped string" when the timer has been stopped:

```
>> t.StopFcn = 'disp(''TIMER stopped.'')';
```

The `TimerFcn` function points to the callback function which is called each time the timer event occurs. This callback function as we already know is `rts_trigger`:

```
>> t.TimerFcn = 'rts_trigger';
```

Then we define a global variable s which will represent a serial interface object. Here we involve the /dev/ttyS0 device:

```
>> s = serial('/dev/ttyS0');
```

The following two statements allow opening/dev/ttyS0 and setting the RTS line to the "on" state:

```
>> fopen(s);
>> s.RequestToSend = 'on';
```

Now we can start the timer and observe the voltage swing on the RTS line using a multimeter. After 15 seconds has passed the RTS line reverses a first time. Each next event will trigger every 5 seconds until the timer elapses 10 times. At that point the timer stops, so we can delete the timer object t and close serial port object s:

```
>> delete(t);
>> clear t;
>> fclose(s);
>> delete(s);
>> clear s;
```

In the same way we can test the DTR line. In this case we have to replace s.RequestToSend with s.DataTerminalReady at every occurrence in the source code.
Reading of the input lines in MATLAB can be best understood with reference to the following example. Here we will check how the state of CTS line changes. In order to implement this experiment let's short circuit together pins 3, 9-10 of the A2 inverter (see Fig.3.30). Then we can set/clear the RTS line and read CTS. It is clear that CTS will be the same as RTS. To build the software type the following source code (Listing 3.15):

Listing 3.15

```
function CTS_test (obj, data)
%UNTITLED Summary of this function goes here
%   Detailed explanation goes here
global s;
i1 = strcmp(s.PinStatus.ClearToSend, 'on');
if (i1)
    s.RequestToSend = 'off';
else
    s.RequestToSend = 'on';
end
```

```
cts = s.PinStatus.ClearToSend;
fprintf(1, ' CTS: %s\n', cts);
end
```

Here we define a `CTS_test` callback function which will be invoked by the timer object `t`. This function must be saved in the `CTS_test.m` file, so that MATLAB can find and execute it.
The following command sequence processes the RTS and CTS signal lines:

```
>> global s;
>> t = timer('StartDelay', 10, 'Period', 5, 'TasksToExecute', 5,
'ExecutionMode', 'fixedRate');
>> t.StartFcn = 'disp(''TIMER started.'')';
>> t.StopFcn = 'disp(''TIMER stopped.'')';
>> s = serial('/dev/ttyS0');
>> fopen(s);
>> s.PinStatus

ans =

    CarrierDetect: 'off'
      ClearToSend: 'off'
     DataSetReady: 'off'
    RingIndicator: 'off'

>> t.TimerFcn = 'CTS_test';
>> start(t);
TIMER started.
 CTS: on
 CTS: off
 CTS: on
 CTS: off
 CTS: on
TIMER stopped.
>> s.PinStatus

ans =

    CarrierDetect: 'off'
      ClearToSend: 'on'
     DataSetReady: 'off'
    RingIndicator: 'off'
```

```
>> fclose(s);
>> delete(t);
>> clear t;
>> clear s;
```

Since we now know how to work with control and status lines in MATLAB, we can build our embedded projects. Let's say that we want to develop a simple embedded system which will control some physical quantity (temperature, pressure, etc.). This system would take the signal from some sensor and switch on/off the control circuit depending on the voltage level of the signal received. The general schematic circuit of such a system is shown in Fig.3.29.

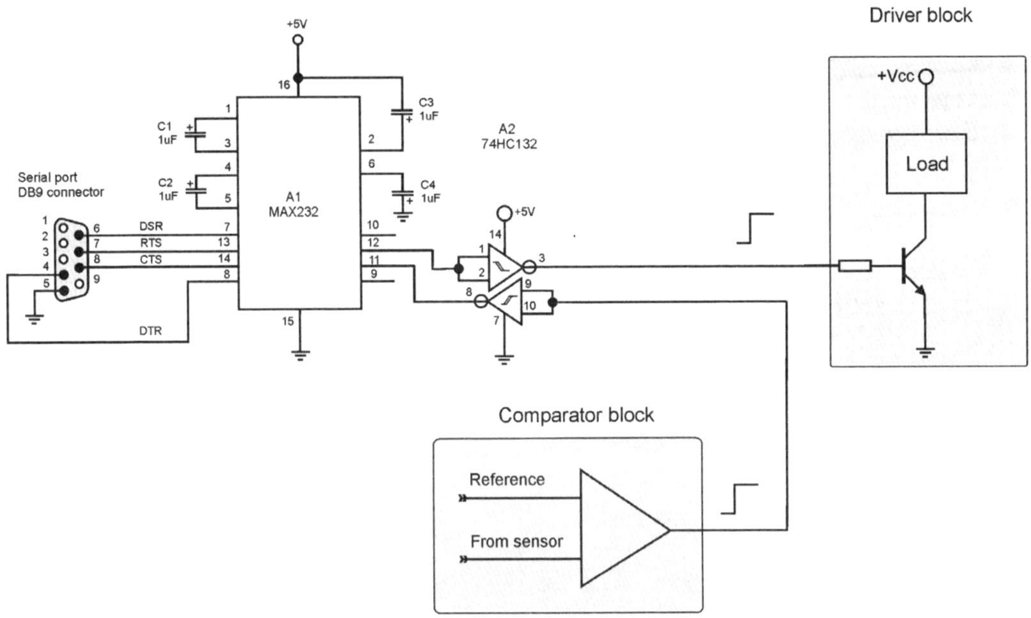

Fig.3.29

As you can see, the signal to estimate is output from a "Comparator block" which may produce one of two possible TTL-compatible signal levels on its output. Let's assume that the comparator's output goes high when a signal from the sensor exceeds the predefined reference voltage and goes low when the sensor's output drops below the reference. The sensor may be any parameter such as temperature, light, pressure or other devices capable of producing some output voltage proportional to the physical parameter being measured (in a similar system I used the LM335 temperature sensor). The signal on the comparator's output may be captured programmatically via the CTS line. The program reads CTS and produces a control signal on the RTS line which drives "Driver block" according to an algorithm determined for this system. By using this general scheme you can easily

realize a variety of embedded systems that may function in a similar way.
The software may be written in a M-file which can be started from a command window of MATLAB. The source code is shown below (Listing 3.16):

Listing 3.16

```
s = serial('/dev/ttyS0');
fopen(s);
for k = 1:7
 res = s.PinStatus.ClearToSend;
 i1 = strcmp(res, 'on');
 if (i1)
    s.RequestToSend = 'off';
 else
   s.RequestToSend = 'on';
 end;
 pause(5);
end;
fclose(s);
delete(s);
clear s;
```

The code is simple and straightforward. At the beginning of the program the serial interface `/dev/ttyS0` is opened by executing the following statements:

```
s = serial('/dev/ttyS0');
fopen(s);
```

Then the program enters a `for()` loop in which the control algorithm is executed. Each iteration begins with a check of the CTS signal line. The state of this line is saved in a `res` variable for further processing:

```
res = s.PinStatus.ClearToSend;
```

The following statement determines the state of CTS:

```
i1 = strcmp(res, 'on');
```

The `i1` variable may be set to logical 1 (true) or logical 0 (false) after `strcmp` has been executed. Being set to 1, `i1` indicates that the CTS line is high; the logical 0 held in `i1` indicates the low level on the CTS line. The following `if...else` statement allows to switch on/off an external load via the

RTS line when `i1` is equal to 0/1. You can easily reverse these conditions by exchange "on" and "off" strings in the `if..else` statement.
The `if.. else` statement is followed by a pause of 5 seconds. After the time interval expires, the next iteration starts. After the seventh iteration has completed the program code leaves the `for()` loop and closes the serial interface opened by the `fopen()` statement. You may change the loop condition or apply other control statements if you want the program to execute continuously.
This example ends our short discussion about the use of the serial port in MATLAB. It is worth mentioning that you can implement any of the projects discussed in this chapter using MATLAB instead of writing a stand-alone C application. Anyway, before you start making your own embedded projects with MATLAB I encourage you to first surf the Internet to acquire lots of other useful documentation about using this particular development tool.

Conclusion

You ascertained that measurement systems controlled via the serial port of the PC may be realized in various ways. There is no so-called 'preferred method' for designing such systems, although you may apply such popular interfaces as SPI and I^2C to provide communications with the PC. Using a SPI interface allows building of relatively simple measurement systems, though you may easily expand your SPI-based system by adding some extra circuitry. The I^2C interface suits when you need to build more complicated systems with multiple devices. You may also apply external PCI modules including two or more serial interfaces that may help you solve most of your problems.

Chapter 4. Sound card in home projects

This chapter describes using a sound card of a PC in your home and laboratory projects. Most applications dealing with a sound card usually perform sound recording and playback. Some advanced applications can also process audio signals, changing their parameters. We will discuss another use of sound cards when the audio output signals actually control external circuits. Before going to real projects let's recall in brief some basics concerning a sound card.
Many times a sound card comes as a built-in device on a motherboard, although there are many professional sound cards implemented as separate standalone modules. These modules are usually applied in high–quality audio processing and may include additional specialized inputs/outputs beyond standard (typical) sound cards. All sound cards provide the basic interfaces as shown in Fig.4.1:

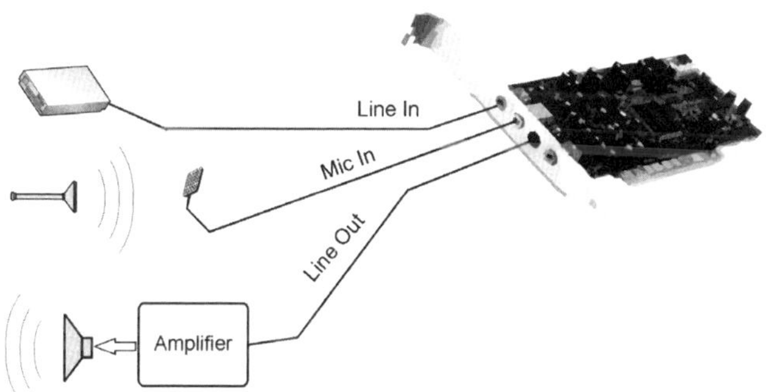

Fig.4.1

Many times the sound card interfaces (ports) are denoted as "Line In", "Mic In" and "Line Out", although you can find other designations as well. The **Line In** port allows listening to or recording from external devices such as digital audio players, microphone preamplifiers or other sources of audio signals. This port can process relatively powerful signals which have sufficient magnitudefor direct processing. Another port of a sound card is the microphone (**Mic In**) port. This interface is intended for capturing relatively weak signals usually arriving directly from microphones or, for example, FM receivers. Sometimes the microphone input may be used to capture signals from special sensors, such as pulse or vibration detectors used in medicine, physics and biology. The common use of this input may be recording notes, lectures, daily personal logs, audio blogs or podcasts. The **Line Out** port provides a preamplified audio signal that is ready for playing via speakers or may be directed to other devices for further processing.

A sound card is usually packaged together with proprietary software and drivers, though Linux operating systems automatically detect and load open source drivers for popular cards. The software interfaces with the sound card allow the user to set parameters, control and check on hardware. Some packages include tools for recording, CD burning, editing or composing.
In order to successfully apply a sound card to embedded projects we need to know how an audio signal is represented in a PC. The original audio signal is analog by its nature and appears as a time-continuous voltage that changes over time. To process such a signal we need to break it into many discrete pieces, and then represent each piece as a binary value. This process is usually called *digitization* and is illustrated in Fig.4.2:

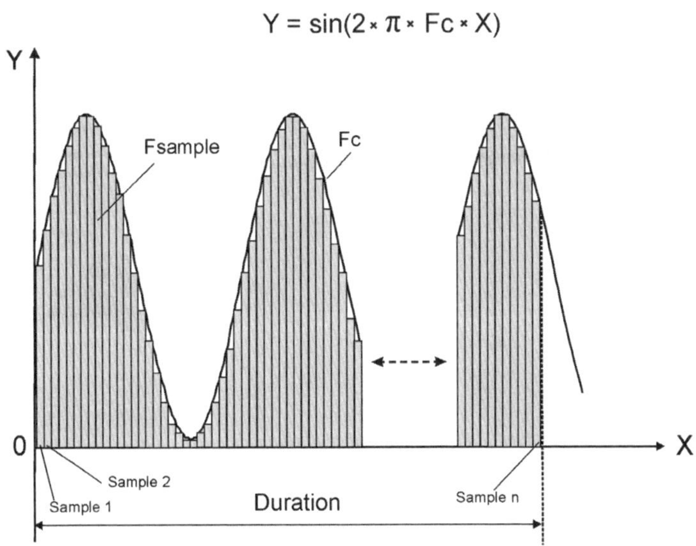

Fig.4.2

It is supposed that a sinusoid voltage signal with a frequency Fc arrives on the input of the sound card (this may be a microphone or a line input). This continuously changing signal is converted into digital samples at some predefined rate. In other words at regularly spaced moments in time the input voltage at that instant is sampled and held constant until the next sample is taken. This reduces the total amount of information to a certain number of discrete voltages. Each sample is stored in some memory buffer allocated for storing this audio data. The size of this buffer must be enough to contain a desired number of samples.
To implement the digitization of an analog signal each sound card contains a built–in analog–to–digital converter (ADC) that captures separate samples and ascribes a binary code to a given magnitude of an input voltage. The particular code is moved to a memory location in the buffer and a pointer is incremented to prepare a place for a next value. Thus, the buffer gets filled with samples gathered over some predefined time interval. Briefly, this is how digitization is implemented.

To play back audio data, we must read the digital samples from a memory buffer and deliver them to a digital–to–analog converter (DAC). The DAC converts each binary code obtained into a voltage and then passes the analog voltage signal to an amplifier to increase the magnitude of the voltage. As you may guess, the sample rate with which the DAC restores a binary code into a continuous analog signal should be equal to that used by the ADC during digitization.
In order for a computer to represent sound accurately, the number of samples must be sufficient that the audio signal can be digitized accurately. For this to happen it has been mathematically proven that the frequency Fsample (see Fig.4.2) of sampling also called the *sampling rate* must be at least twice as high as the highest possible frequency Fc of the input signal. As the high limit of a sound signal is about 20KHz, the sampling rate Fsample must be at least 40KHz or even higher. The standard for recording sound is 44100 samples per second, although the most modern audio devices can operate at higher sampling rates. After digitization has been completed the audio data located in a buffer may be saved in some file having a special format known to audio applications (.wav or another) or may be played back producing a sound via speakers.
A sound card may be successfully applied to various home and laboratory projects. Possible applications include processing low-level, low-frequency input signals from various sources, conversion/coding/decoding, mixing and modulating signals, generating waveforms for testing and measurement purposes and so on. The sound card may even be successfully used in control systems capable of triggering multiple loads. Using the sound card in embedded projects involves both hardware and software, so we will discuss both aspects of design. Let's start from learning how the users application software may operate with sound in the Linux environment.
The program interacts with a sound card using specific software interfaces; each of them basically includes the kernel and the user parts. A user part includes a set of API calls through which the users application talks to the specific kernel driver. The configuration of the kernel part of a sound interface depends on the sound system used in Linux. We will consider in brief two of the most popular sound systems: OSS (Open Sound System) and ALSA (Advanced Linux Sound Architecture).
The Open Sound System is based on using standard UNIX devices which are accessed through standard POSIX API calls such as `read`, `write`, `ioctl`, etc. OSS aims to write sound-based applications that work with any sound control hardware, even though the hardware interface varies greatly from one type to another. This sound system is deprecated because it doesn't allow low-level sound processing and configuring. Nowadays some modern distributions of Linux have limited support for OSS, preferring instead to use the ALSA sound system. Nevertheless, the OSS sound system is still used because of its simplicity and clearness in the cases when the developer needs to operate with sound using only basic POSIX API calls. OSS may operate with `dev/mixer`, `/dev/sndstat`, `/dev/dsp`, `/dev/audio`, `/dev/sequencer` and `/dev/music` devices.
Let's discuss OSS first. Once we deal with the given sound system we must know what OSS sound devices are presented and configured in the Linux system.

First of all we have to be confident that Linux discovered our sound card and properly configured it. The common way to check the presence of the device is to examine a `/dev/sndstat` device that is the simplest one provided by the sound driver. This is a read-only device file which only reports

information about the sound driver in human-readable form; we cannot configure a given sound device in some way. This device is useful when checking hardware configuration (DMA channel, IRQ number, etc.). We can also get the version of the sound driver being used.
If the sound driver is absent, opening `/dev/sndstat` causes an `ENODEV` error to be returned. If the device was successfully opened we get a lot of information about sound the subsystem. Below you can see information about the sound card installed in my PC:
NOTE: For some modern Linux distributions such as fedora, there is no /dev/sndstat. If a different sound server is used

it must be clarified that ALSA is used and not OSS(or pulseaudio).

```
$ cat /dev/sndstat

Sound Driver:3.8.1a-980706 (ALSA v1.0.20 emulation code)

Kernel: Linux yury-desktop 2.6.31-16-generic #53-Ubuntu SMP Tue Dec 8
04:01:29 UTC 2009 i686

Config options: 0
Installed drivers:
Type 10: ALSA emulation
Card config:

ESS ES1938 (Solo-1) rev 0, irq 17

Audio devices:
0: ESS Solo-1 (DUPLEX)

Synth devices: NOT ENABLED IN CONFIG
Midi devices:

0: ESS ES1938 (Solo-1) MIDI

Timers:
31: system timer

Mixers:
0: ESS Solo-1
```

The same result might be achieved by opening a `/dev/dsp` device. A sound card mostly offers three main features:

- Digital sample input/output;
- Frequency modulation output;
- A midi interface;

Each of these features is provided through its own device driver interface. For digital samples it will be `/dev/dsp`; the frequency modulation is processed via `/dev/sequencer` and the MIDI interface will be provided through `/dev/midi`. The sound settings (volume, balance, etc.) are controlled via the `/dev/mixer` interface.

Although the above `cat` command allows you to read the configuration of the sound device, refrain from writing any code based on this information because it describes general features of your device. This information may be changed over time, such as after the sound driver has been enhanced. In practice we must deal with a `/dev/dsp` device when programming a sound card.
As was mentioned, nowadays pretty much all Linux operating systems support the ALSA sound system, so we will discuss its programming architecture and features. Before dealing with ALSA it is worth checking if your sound card has an appropriate ALSA driver. All supported sound card chips are enumerated on the ALSA web site www.alsa-project.org together with the appropriate drivers.
There are two preferred application approaches to use a sound card to our home projects.. The first; we can use it to produce test waveforms (sinusoid, square wave, triangle, sawtooth, etc.). Those signals surely come in handy when tuning up various circuits. Thus, use of the sound card allows us to quickly build an almost costless, but very useful signal generator. We may further consider at least two methods to fulfill this approach.

We can realize such a signal generator entirely in software written in say, the C programming language. In this case we must calculate the waveforms we want by using some math functions accessible from the `math.h` library. In this way we can build the signal of interest mathematically, then calculate samples, fill a memory buffer with these samples and, finally, play the sound. This approach doesn't require designing any circuitry except for perhaps introducing some external signal buffer stage on the Line-out output of the sound card. This could be a very simple circuit., For example, a voltage follower on a wide-band opamp – that isolates the sound card output from the target device(s). In this way we can get an extremely low-cost, but quite good performance low-frequency signal generator which forms signals of predetermined waveforms using the DAC of the sound card.

Another approach to generate signals is to create a basic signal, usually sinusoid, programmatically and then apply some additional circuitry to convert the given basic signal, say, into a square wave, sawtooth or other waveform. The benefit of this method is that we can gain several signals at a time – the basic signal on the Line Out output and additional signal waveforms somewhere in external circuitry.
Let's see how to manipulate audio signals in both OSS and ALSA sound systems.

Programming sound with OSS

Before attempting to program a sound device using OSS let's revisit `/dev/dsp` in detail. The `/dev/dsp` device appears in Linux as the digital sampling and digital recording device. This device is also responsible for implementing multimedia applications. Specifically, when the program writes to that device it causes the digital–to–analog converter (DAC) to produce sound. Again, reading the device activates the analog–to–digital converter (ADC) that starts sound recording and analysis of the result. The name DSP stands for "Digital Signal Processor" which is a specialized processor chip optimized for digital signal processing and analysis; modern sound cards usually have a built-in DSP chip dedicated to sound processing. Some sounds cards may provide more than one digital sampling device. For example, a second device will be available as `/dev/dsp1`. Unless stated otherwise, this device operates in the same manner as `/dev/dsp`.

In fact, the DSP device really includes two devices, the DAC and ADC. Opening for read-only access allows you to use the analog–to–digital converter for sound input. Opening for write only will access the D/A converter for sound output.Generally speaking, the program should open the device either for read only or for write only. It is possible to perform both read and write operations on the device at a time, although with some restrictions. Note that only one process can have the DSP device open at a time; another process attempting to reopen it will fail with an error code of `EBUSY`.

Reading from the DSP device returns digital sound samples obtained from the A/D converter. Analog data is converted into digital samples by the analog to digital converter under control of the kernel sound driver and stored in a buffer internal to the kernel. When an application program invokes the `read()` system call, the data is transferred to the calling program's data buffer. It is important to understand that the sampling rate is dependent on the kernel driver, and not the speed at which the application program reads it.When reading from `/dev/dsp` you will never encounter an end–of–file (EOF) condition. If data is read too slowly, less than the sampling rate, excess data will be discarded resulting in gaps in the digitized sound. If you read the device too quickly, the kernel sound driver will block your process until the required amount of data is available.

The input source which is used for audio data reading depends on the mixer setting – the default is the microphone input. The format of the digitized data depends on what parameters were taken by the `ioctl()` call when configuring an audio device. Each time the device is opened, its parameters are set to default values. The default parameters are as follows: 8-bit unsigned sample, one channel (mono) and an 8 KHz sampling rate.

Writing a sequence of digital samples to the DSP device, as you already know, produces a sound. Again, the format of output audio signal is set by invoking the `ioctl()` calls with appropriate parameters, but default settings are the same as those defined for input signals (8-bit unsigned data, mono and 8KHz sampling rate).

If the data is being written too slowly, there will be dropouts or pauses in the sound output. Writing the data faster than the sampling rate will simply cause the kernel sound driver to block the calling process until the sound card hardware is ready to process the new data. As in the case of capturing input data, you can change the default parameters through `ioctl()` calls. In general, you should set the parameters after opening the device, but before any read/write operations. Note, that you should also set the parameters in the order in which they are listed below.

If a parameter is out of range, the driver will set it to the closest value. For example, attempting to use 16-bit sampling with an 8-bit sound card will result in the driver selecting 8 bits, but no error will be returned. It would be worth verifying that the value returned is acceptable to your application. All of the `ioctl()` calls for the DSP device take the parameters with names starting with `SOUND_PCM`. Parameters that are similar to `SOUND_PCM_READ_XXX` are used to return just the current value of a parameter. To change the values, the `ioctl()` calls should take the parameters that may look like `SOUND_PCM_WRITE_XXX`. As discussed above, these calls also return the selected value, which is not necessarily the same as the value passed to the sound driver.

The `ioctl()` constants are defined in the header file `linux/soundcard.h`. Let's examine each of them in detail.

- **`SOUND_PCM_WRITE_BITS`** sets the sample size, in bits. Valid choices are 8 and 16, but note that some cards do not support 16;
- `SOUND_PCM_READ_BITS` returns the current sample size, which should be either 8 or 16 bits;
- `SOUND_PCM_WRITE_CHANNELS` sets the number of channels -- 1 for mono, 2 for stereo. When running in stereo mode, the data is interleaved when read or written, in the format left–right–left–right. Note that some sound cards do not support stereo mode, so you must check the actual number of channels returned by the call;
- `SOUND_PCM_READ_CHANNELS` returns the current number of channels, either 1 or 2;
- `SOUND_PCM_WRITE_RATE` sets the sampling rate in samples per second. Remember that all sound cards have a limit on the range; the driver will round the rate to the nearest speed supported by the hardware, returning the actual (rounded) rate. Typical lower limits are 4 KHz; upper limits are 13, 15, 22, or 44 KHz;
- `SOUND_PCM_READ_RATE` returns just the current sampling rate. This is the rate used by the kernel, which may not be exactly the rate given in a previous call to `SOUND_PCM_WRITE_RATE`, because of the previously discussed rounding.

Now we will consider a very basic example of programming the DSP device. The program will create a sound of 1000 Hz using the `/dev/dsp` device. The sound will last for 5 seconds. You can see a source code below (Listing 4.1):

Listing 4.1

```
#include <unistd.h>
#include <fcntl.h>
#include <sys/types.h>
#include <sys/ioctl.h>
#include <stdlib.h>
#include <stdio.h>
#include <linux/soundcard.h>
#include <math.h>

#define FREQUENCY 1000     // frequency = 1000Hz
#define DURATION  5        // how many seconds of speech to store
#define RATE      11025    // the sampling rate
#define SIZE      8        // sample size: 8 or 16 bits
#define CHANNELS  1        // 1 = mono 2 = stereo


int main()

{
  int fd;                // sound device file descriptor
  int param;             // argument for ioctl calls
  int status;            // return status of system calls

  int i1;
  int n1;
  n1 = DURATION * RATE;

 // this buffer holds the digitized audio

  char buf[n1];
  double x;

  // open sound device

  fd = open("/dev/dsp", O_RDWR);
  if (fd < 0)
  {
```

```
    perror("open of /dev/dsp failed");
    exit(1);
  }

  // set sampling parameters

  param = SIZE;       // sample size
  status = ioctl(fd, SOUND_PCM_WRITE_BITS, &param);
  if (status == -1)
   perror("SOUND_PCM_WRITE_BITS ioctl failed");
  if (arg != SIZE)
   perror("unable to set sample size");

  param = CHANNELS; // mono or stereo
  status = ioctl(fd, SOUND_PCM_WRITE_CHANNELS, &param);
  if (status == -1)
   perror("SOUND_PCM_WRITE_CHANNELS ioctl failed");
  if (arg != CHANNELS)
    perror("unable to set number of channels");

 param = RATE;       // sampling rate
 status = ioctl(fd, SOUND_PCM_WRITE_RATE, &param);
 if (status == -1)
   perror("SOUND_PCM_WRITE_WRITE ioctl failed");

// filling the buffer with samples

for (i1 = 0; i1 < n1; i1++)
{
 x = sin (i1 * 2.0 * M_PI * (double) FREQUENCY / (double) RATE);
 buf[i1] = (char)(127*x + 128);
}

status = write(fd, buf, sizeof(buf)); // play sound buffer
if (status != sizeof(buf))
 perror("wrote wrong number of bytes");
status = ioctl(fd, SOUND_PCM_SYNC, 0);
if (status == -1)
 perror("SOUND_PCM_SYNC ioctl failed");
close(fd);
return 0;
}
```

Note that we create the sinusoid programmatically, so we need to include a `math.h` header file where this function is defined. Also we need (this is specifically for `gcc`) to compile this source with the following command:

```
$ gcc playsound.c /usr/lib/libm.a -o playsound
```

Here the `playsound.c` is a file where we saved the source code. We need to mention the `/usr/lib/libm.a` in the compilation string above, otherwise we get an error while compiling a source.

While running, this program yields a 1000Hz tone which sounds for 5 seconds. To make things clear let's discuss what the application does. Since the program deals with numerous samples there will be a memory buffer where audio data is held. This buffer is defined as `char buf[n1]` and its size `n1` is determined as `DURATION * RATE`. To understand this, please, refer to Fig.4.2.

The `RATE` constant determines the sampling rate that, in our case, is selected to be equal to 11025 samples per second. The `DURATION` defines the time which we take for sounding. In our case we take 5 s to play the sinusoid signal. As the audio device is some kind of file we must open it before doing something:

```
fd = open("/dev/dsp", O_RDWR);
```

After a system call `open()` has succeeded the `fd` variable will hold a descriptor used in read/write operations. The statements that follow are needed for configuring the audio parameters. As you can guess, settings of audio parameters are implemented through an `ioctl()` call which takes the appropriate constants as parameters. The following statements will set the parameters for an audio device (sample size 8 bit, mono mode and sampling rate 11025):

```
. . .
param = SIZE;
status = ioctl(fd, SOUND_PCM_WRITE_BITS, &param);
. . .
param = CHANNELS;
status = ioctl(fd, SOUND_PCM_WRITE_CHANNELS, &param);
. . .
param = RATE;
status = ioctl(fd, SOUND_PCM_WRITE_RATE, &param);
```

Note that the parameters selected are valid for a sound card installed in my desktop PC. You should carefully consult documentation on your own sound card, in order that you may configure your device properly. Some parameters, specifically sample rate, may differ for various sound cards. If you are not certain about configuring your sound card, you may select common parameters applicable for most audio devices. For example, you could select an 8-bit sample size, sample frequency of

8000 KHz and a single channel to capture and play. This works for most sound cards.
Let's go back to our source code. As our program produces a 1000Hz tone for 5 s we need to fill the buffer `buf` with appropriate values. This task is accomplished inside a `for()` loop that repeats `n1` times. Here the variable `x` of a double type will hold the current value of a sinusoidal curve corresponding to a particular `i1` sample. The following statement calculates the value of a sinusoid for `i1` sample and saves the result in the `x` variable:

```
x = sin (i1 * 2.0 * M_PI * (double) FREQUENCY / (double) RATE);
```

This formula becomes clear when we recall how a sinusoid curve is represented in math. In the simplest case, a sinusoid signal may be expressed as follows:

$y = \sin(2\pi ft)$,

where `f` is a frequency (`Fc` in Fig.4.2) and `t` is a time interval. In our case we need to jump from a `t` variable to a value, depending on the current sample. This value may be calculated as `i1` divided by `RATE`. After `x` has been calculated its value should be adjusted according to an 8-bit unsigned char. This is achieved using the following statement:

```
buf[i1] = (char)(127*x + 128);
```

Thus after the `for()` loop exits, the buffer `buf` will contain all `n1` samples recalculated as the 8-bit values. After filling the buffer we can write it to the `/dev/dsp` audio device by invoking a `write()` function:

```
status = write(fd, buf, sizeof(buf));
```

As some time is required to complete playing, the following statement should be executed:

```
status = ioctl(fd, SOUND_PCM_SYNC, 0);
```

Finally, we need to close a file descriptor:

```
close(fd);
```

When testing this application make sure that you connect a signal source to the appropriate audio input (microphone or line). Also check the default input for your sound card and change it if needed.
Another task you may want to accomplish is to capture (read) sound coming from some external source (microphone, amplifier, line, etc.) programmatically and play the saved audio data via speakers. Below you can see a C source code of this application (Listing 4.2):

Listing 4.2

```
#include <unistd.h>
#include <fcntl.h>
#include <sys/types.h>
#include <sys/ioctl.h>
#include <stdlib.h>
#include <stdio.h>
#include <string.h>
#include <linux/soundcard.h>

#include <math.h>

#define DURATION  5       // how many seconds of (speech) sound to
store
#define RATE      11025   // the sampling rate
#define SIZE      16      // sample size: 8 or 16 bits
#define CHANNELS  2       // 1 = mono 2 = stereo


int main()
{
  int fd;         // sound device file descriptor
  int param;      // argument for ioctl calls
  int status;     // return status of system calls

  int n1;

  // open sound device

  n1 = DURATION * RATE * CHANNELS; // size of memory buffer
  char buf[n1];                    // this buffer holds a digital
                                      audio data

  fd = open("/dev/dsp", O_RDWR);
  if (fd < 0)
   {
    perror("open of /dev/dsp failed");
    exit(1);
   }
```

```
// set sampling parameters

param = SIZE;                      // sample size
status = ioctl(fd, SOUND_PCM_WRITE_BITS, &param);
if (status == -1)
  perror("SOUND_PCM_WRITE_BITS ioctl failed");
if (arg != SIZE)
  perror("unable to set sample size");

param = CHANNELS; // mono or stereo
status = ioctl(fd, SOUND_PCM_WRITE_CHANNELS, &param);
if (status == -1)
  perror("SOUND_PCM_WRITE_CHANNELS ioctl failed");
if (arg != CHANNELS)
  perror("unable to set number of channels");

param = RATE;       // sampling rate
status = ioctl(fd, SOUND_PCM_WRITE_RATE, &param);
if (status == -1)
  perror("SOUND_PCM_WRITE_WRITE ioctl failed");

while (1)
 {
  // loop until Control-C

  printf("Say something:\n");
  status = read(fd, buf, sizeof(buf)); // record some sound
  if (status != sizeof(buf))
    perror("read wrong number of bytes");
  printf("Received:\n");

  status = write(fd, buf, sizeof(buf)); // play it back
  if (status != sizeof(buf))
    perror("wrote wrong number of bytes");

  // wait for playback to complete before recording again

  status = ioctl(fd, SOUND_PCM_SYNC, 0);
  if (status == -1)
   perror("SOUND_PCM_SYNC ioctl failed");
```

```
    }
    close(fd);
    return 0;
}
```

Let's dig into the source code. As usual, we must allocate a memory buffer for the audio data. The size of the buffer (a `buf` variable) is held in the `n1` variable. The program opens the `/dev/dsp` device and configures its parameters using an `ioctl()` function. After that the program enters an endless `while()` loop, where data is read from the input and then is delivered to the output. Reading of data is accomplished by using a `read()` call and writing is accomplished by invoking the `write()` function.

Save this source code in the `record_play_sound.c` file and compile it using the following command string:

```
$ gcc record_play_sound.c /usr/lib/libm.a -o record_play_sound
```

After successful compilation you can launch the program. Once started, the program waits while you pronounce some phrase and then replays the audio buffer you just saved.

These two software projects may serve as a basis for programming and processing tones and various user-defined signals. For example, when you want to build your very low-cost signal generator, you may programmatically generate a wide-band sinusoid signal, and then convert it to a square-wave, triangle, saw-tooth, etc. by introducing some simple external circuits attached to the sound card output. You can also control some external loads using tone-decoders; each configured to sense a particular frequency.

As the input audio signals range from about 20Hz through to 20000 Hz, that allows us to capture signals from various transducers that convert some physical quantity (temperature, light, pressure, etc.) into frequency. Such signals after analyzing may be used in control algorithms. Popular embedded projects also include voice-controlled devices. Nearly all embedded projects based on a sound card require some hardware interface, and we will see how we can build such interfaces.

Programming sound with ALSA

Programming sound using ALSAs capabilities gives the developer key advantages over OSS. With ALSA you are allowed to configure most audio parameters and that provides you greater flexibility and creative power while processing sound signals. In addition, the developer gains a great degree of control over audio processing from their applications.

If you want to work with the ALSA sound system you should be aware that ALSA is configured in the system's kernel. After ALSA has been installed, the `/proc/asound` directory is created, so you can simply check for its presence in a file system. Once configured the ALSA sound system allows operating of the sound card using ALSA API calls.

If you don't find asound for some reason, you can download and configure ALSA by yourself. The latest version of the ALSA sound system software may be found on www.alsa-project.org (I used alsalib-1.0.23). After downloading and unpacking this library you should install it by entering the following commands:

```
configure
make install
```

When you have successfully configured the ALSA system, the very next thing you should do is to check if information on your sound card(s) installed is available. To do so you can view the contents of the /proc/asound/cards file. Each entry there corresponds to a (real or simulated) sound card. Below is what I saw in my Linux system after ALSA had been configured:

```
yury@yury-desktop:~$ ls -l /proc/asound/cards

-r--r--r-- 1 root root 0 2010-04-24 18:55 /proc/asound/cards

yury@yury-desktop:~$ cat /proc/asound/cards

 0 [ICH5]: ICH4 - Intel ICH5

                  Intel ICH5 with ALC650F at irq 17
```

As you can see there is a single sound card (specified as card 0) present in the system.
Note that the library API functions work with logical device names rather than device files. The device names may be either real hardware devices or plug-ins. Hardware devices are usually presented as hw:i,j, where i is the card number and j is the device on that card. From this point the first sound device appears as hw:0,0. The alias "default" refers to the first sound device and is used in the examples below.
Let's go to the practice. The following C source code presents the ALSA program that will generate a sinusoid waveform (Listing 4.3):

Listing 4.3

```
#include <stdio.h>
#include <stdlib.h>
#include <alsa/asoundlib.h>
#include <math.h>

main (void)
{
```

```
  int i;
 int err;
 double x;
 double frequency = 500;
 unsigned int rate = 44100;

 short buf[2048];
 snd_pcm_t *phandle;
 snd_pcm_hw_params_t *hw_params;

 if ((err = snd_pcm_open (&phandle, "default", SND_PCM_STREAM_
PLAYBACK, 0)) < 0)
 {
   printf ("Cannot open audio device\n");
   exit (1);
 }

 if ((err = snd_pcm_hw_params_malloc (&hw_params)) < 0)
  {
   printf ("Cannot allocate hardware parameter structure\n");
   exit (1);
  }

 if ((err = snd_pcm_hw_params_any (phandle, hw_params)) < 0)
  {
   printf ("Cannot initialize hardware parameter structure\n");
   exit (1);
  }

 if ((err = snd_pcm_hw_params_set_access (phandle,  hw_params,
SND_PCM_ACCESS_RW_INTERLEAVED)) < 0)
  {
   printf ("Cannot set access type\n");
   exit (1);
    }

 if ((err = snd_pcm_hw_params_set_format (phandle, hw_params, SND_
PCM_FORMAT_S16_LE)) < 0)
  {
    printf ("Cannot set sample format\n");
   exit (1);
```

```
 }

if ((err = snd_pcm_hw_params_set_rate_near (phandle, hw_params,
&rate, 0)) < 0)
 {
  printf ("Cannot set sample rate\n");
  exit (1);
 }

if ((err = snd_pcm_hw_params_set_channels (phandle, hw_params,
1)) < 0)
 {
  printf ("Cannot set channel count\n");
  exit (1);
 }

if ((err = snd_pcm_hw_params (phandle, hw_params)) < 0)
 {
  printf ("Cannot set parameters\n");
  exit (1);
 }

snd_pcm_hw_params_free (hw_params);

if ((err = snd_pcm_prepare (phandle)) < 0)
 {
  printf ("Cannot prepare audio interface for use\n");
  exit (1);
 }

for (i = 0; i < 2048; i++)
 {
   x = sin(i * 2.0 * M_PI * frequency / (double)rate);
   buf[i] = (short)(32767 * x + 32768);
  }

for (i = 0; i < 50; ++i)
{
  if ((err = snd_pcm_writei (phandle, buf, 2048)) != 2048)
   {
    printf ("Writing to audio interface failed\n");
```

```
      exit (1);
    }
  }

 snd_pcm_close (phandle);
 exit (0);
}
```

This application generates a sinusoidal signal which is expressed as sin(2πF), where F is a frequency. To provide an access to the ALSA API functions we need to place an `include` directive with the appropriate header file:

```
#include <alsa/asoundlib.h>
```

In this program we defined some variables. The `frequency` variable of a double type contains the value of our frequency (500 Hz in our case). The `rate` variable of an int type indicates a sampling rate that we will take equal to 44100 samples/s. Also we need to allocate some buffer to hold audio data for the playback. That will be a `buf` variable where the program will move samples. We are going to form 16-bit samples, so `buf` will contain 2-byte (short) values. The size of `buf` is selected as 2048, although you may select another value. The `x` variable of a double type will hold the discrete values of a sinusoid taken in the `for()` loop.
In order to set the hardware parameters for the stream, we need to allocate a `hw_params` variable of type `snd_pcm_hw_params_t`. A `phandle` variable will contain the handle to the sound device and will be used in subsequent calls to manipulate the sound stream.
First our program must open the audio interface and get its handle for further manipulations. The call to `snd_pcm_open` opens the default audio device and sets the access mode to `PLAYBACK`. This function subsequently returns a handle in its first argument that points to `phandle`. After this we should initialize the `hw_params` structure with the audio parameters required. The call

```
snd_pcm_hw_params_malloc (&hw_params)
```

reserves the buffer to hold these parameters. The `snd_pcm_hw_params_any (phandle, hw_params)` function that follows initializes the `hw_params` structure with default values. After that we can change some basic values stored in `hw_params`. First, our audio data may be interleaved, so we configure that mode by executing the `snd_pcm_hw_params_set_access (phandle,hw_params,SND_PCM_ACCESS_RW_INTERLEAVED)` function. The following several calls set up parameters (audio data format, sampling rate and number of channels) that we have selected for audio processing:

```
snd_pcm_hw_params_set_format (phandle, hw_params, SND_PCM_FORMAT_
S16_LE)
```

```
snd_pcm_hw_params_set_rate_near (phandle, hw_params, &rate, 0)
snd_pcm_hw_params_set_channels (phandle, hw_params, 1)
```

Once complete the program writes this configuration to the device driver using the following function:

```
snd_pcm_hw_params (phandle, hw_params)
```

After that we can free the `hw_params` structure by calling the `snd_pcm_hw_params_free (hw_params)` function and prepare to launch the sound device (the `snd_pcm_prepare (phandle)` function).
Then we should fill `buf` with the samples. The `for()` loop does that:

```
for (i = 0; i < 2048; i++)
 {
  x = sin(i * 2.0 * M_PI * frequency / (double)rate);
  buf[i] = (short)(32767 * x + 32768);
 }
```

First we calculate the `sin` function for each of the 2048 samples that will be stored as 16-bit values in `buf`. Each value of the `x` variable is scaled to a short value in order to provide a 16-bit sound reproduction
A following `for()` loop plays the buffer 50 times using the `snd_pcm_writei (phandle, buf, 2048)` function. After the sound reproduction completes we close the handle by calling `snd_pcm_close (phandle)`.
As you can see the sound programming with ALSA is actually not as difficult as it may seem. You can easily modify this source code in order to generate more complex signals, for example, programmatically controlled AM or PWM, etc.
To compile and build the executable we can save this source code in the `alsaexample1.c` file and execute the following command:

```
$ gcc -o alsaexample1 alsaexample1.c -lasound
```

Using the ALSA sound library allows the developer to select and modify the various parameters to tailor the desired characteristics of a sound signal. For example, to reach a better accuracy we can slightly modify the above listing by changing the audio data format from signed 16-bit to a 32-bit float value. In this case we can replace `SND_PCM_FORMAT_S16_LE` in the `snd_pcm_hw_params_set_format` function on `SND_PCM_FORMAT_FLOAT_LE`. This provides a better resolution that, in turn, leads to a better waveform quality. Such a modification also simplifies the `for()` loop which fills the buffer with samples:

```
for (i = 0; i < 2048; i++)
     x[i] = (float) sin(i * 2.0 * M_PI * frequency / (double)rate);
```

Sometimes you may want to playback some waveforms without much coding. This can be accomplished by using a convenient API call snd_pcm_set_params(). This function allows accepting of the snd_pcm_hw_params_t structure and all. The following source code illustrates that (Listing 4.4):

Listing 4.4

```
#include <stdio.h>
#include <stdlib.h>
#include <alsa/asoundlib.h>
#include <math.h>

main (void)
{
 int i;
 float x[2048];
 double frequency = 1000;
 unsigned int rate = 44100;

 int err;
 short buf[2048];
 snd_pcm_t *phandle;


 if ((err = snd_pcm_open (&phandle, "default", SND_PCM_STREAM_
PLAYBACK, 0)) < 0)
   {
    printf ("Cannot open audio device\n");
    exit (1);
   }

 if ((err = snd_pcm_set_params(phandle,                          //
device handle
                               SND_PCM_FORMAT_FLOAT_LE,          //
                               set PCM format
                               SND_PCM_ACCESS_RW_INTERLEAVED,//
                               set access
                               1,                                // 1
```

```
                                   channel
                                   rate,                          //
                                   sampling rate
                                   1,                             //
                                   allow resampling
                                   500000)) < 0)                  //
latency
  {
   printf ("Cannot set parameters)\n");
   exit (1);
   }

 if ((err = snd_pcm_prepare (phandle)) < 0)
  {
   printf ("Cannot prepare audio interface for use\n");
   exit (1);
  }

 for (i = 0; i < 2048; i++)
  {
   x[i] = (float) sin(i * 2.0 * M_PI * frequency / (double)rate);
  }

 for (i = 0; i < 50; ++i)
  {
   if ((err = snd_pcm_writei (phandle, x, 2048)) != 2048)
    {
     printf ("Writing to audio interface failed\n");
     exit (1);
   }
  }

 snd_pcm_close (phandle);
 exit (0);
}
```

In order to read the input audio data we must make some changes to our previous listing. First we need to open the audio interface for capturing data by indicating the `SND_PCM_STREAM_CAPTURE` value in the `snd_pcm_open` function. Also we must introduce the `snd_pcm_readi` function for continuous reading audio data in some `for()` loop; this function takes the same parameters as `snd_pcm_writei`.

Interfacing a sound card to external circuits

When developing hardware interfaces to our sound card we must know the electrical characteristics of the audio signal. The magnitudes of those signals must be kept within specified levels; usually, for input signals the peak voltage should not be higher than 1V. Higher levels may damage your sound card, so you need to assemble simple circuitry to limit the upper level of input signals. We may accept this level to be about 1V or even less. Additionally we must be sure that an input signal is free of high-frequency noise, so it would be worth introducing an active low-pass filter between the signal source and the sound card line input. A possible solution is shown in the schematic circuit of Fig.4.3.

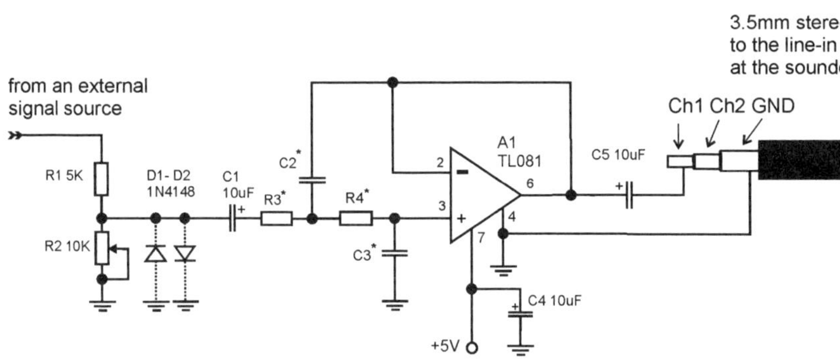

Fig.4.3

In this circuit the input signal from an external source enters the single channel Ch1 through the voltage divider R1–R2 that allows you to lower the voltage level simply by adjusting the value of the variable resistor R2. Protection diodes D1–D2 further restrict this level to a safe value. The diodes should operate with audio frequencies, of tens of kilohertz, so we should select high-speed parts such as 1N4148/1N4448 or similar. The input signal is delivered to the active low-pass filter on TL081 opamp (A1) via capacitor C1. As you can see from Fig.4.3, the input and output signals of this circuit are AC coupled through capacitors C1 and C5 having sufficiently high values – this lessens the cut-off frequency of the circuit. Usually the input signal is also AC coupled to the sound-card itself.

The output audio signal may be a basis for designing various signal generators. Using simple circuitry we can gain further signals from a single audio output. One possible approach which I use in my home laboratory is based on the following circuit (Fig.4.4).

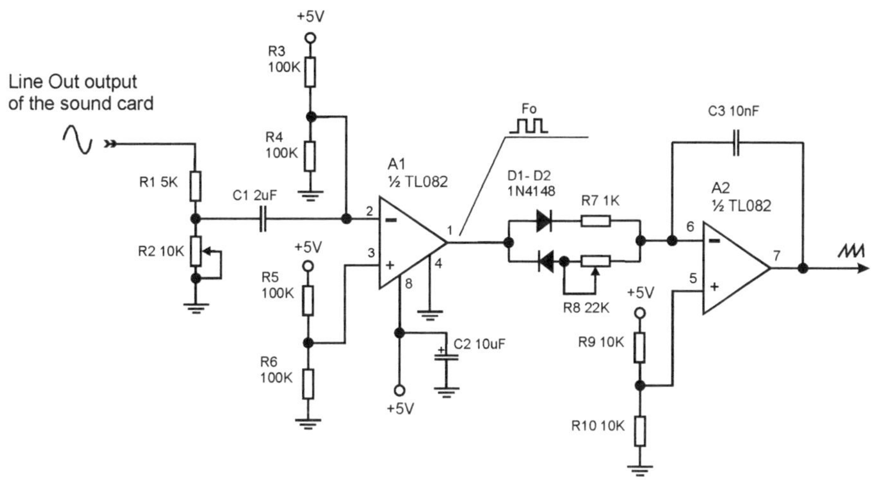

Fig.4.4

This circuit, in fact, consists of two sub-circuits or stages. The first stage is built using the opamp A1 configured as a comparator. This comparators output swings from low to high depending on the level of an AC signal on the inverting input (pin 2). We take the sinusoid signal from our sound card, deliver it to the inverting input of A1 and get a series of pulses on the comparator's output (pin 1). These pulses have the same frequency as the input signal on the inverting input of A1. Those pulses go to the second stage (opamp A2) that functions as an integrator. The output of A2 (pin 7) yields a sawtooth signal. The capacitor C3 charges and discharges through separate chains (D1-R7 and D2-R8) which allow separate configuring of the rising and falling edge timings of a sawtooth signal on pin 7 of A2. The resistor R8 allows setting the slope of the waveform. The values of R7, R8 and C3 are selected to obtain the best performance and linearity in the range of about 1 to 3 KHz. You may experiment with these values while configuring this circuit. To further improve the circuit you may replacing the D1-D2 diodes using Shottky diodes. In general, this circuit operates well over a wide band of input frequencies.

The output audio signal may be successfully applied to control multiple loads using tone–decoder circuits. One channel of such a circuit is shown in Fig.4.5:

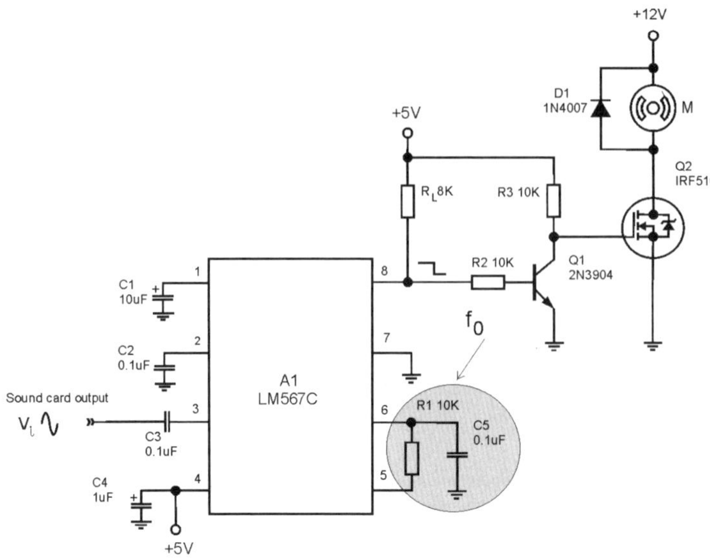

$$f_0 = \frac{1}{1.1 \times R1 \times C5} \qquad BW = 1070 \times \sqrt{\frac{V_i}{f_0 \times C2}} \text{ in \% of } f_0$$

Fig.4.5

Here we use a popular chip LM567C. The LM567C is a tone–decoder which provides a low-level voltage level on its output (pin 8) when an input signal (V*i* in Fig.4.5) falls into the given pass–band. The tone–decoder circuit includes a voltage controlled oscillator with center frequency f_0 defined as shown in Fig.4.5. The bandwidth (BW) of the tone–decoder is dependent on the central frequency f_0 and the capacitor C2.

External components R1–C5 and C2 are used to independently set the center frequency, bandwidth and output delay. The built–in oscillator of the LM567C runs at a frequency determined by the R1–C5 pair attached to pins 5-6. The external frequency produced by a sound card passes onto pin 3 via capacitor C3. If this frequency becomes equal to that of the built–in oscillator, the output of the tone–decoder (pin 8) goes low. If the external frequency is beyond BW the output level on pin 8 goes high. In our case the center frequency f_0 will be about 900 Hz. If an input frequency is equal to the center frequency f_0, the output of the tone–decoder goes low. The input frequency beyond the pass-band results in raising the voltage level on pin 8.

Thus, by changing the frequency on the LM567C's input we can switch on/off some load connected to pin 8 of A1. The output of tone–decoder yields TTL–compatible levels and can drive a load up to 100 mA. To provide a control for For heavy loads you would add an additional stage of power transistor (bipolar or MOSFET). A typical example would use a stage with a POWER MOSFET IRF510 (Q2) driven by the output of A1 via bipolar transistor 2N3904 (Q1).

You may develop another configuration for your own output stage by using, for instance, a single Darlington transistor TIP127 driven directly from pin 8 of LM567C as well.

If you want to drive some AC load you can apply, for example, an optotriac (MOC3023, 3061, 3063, etc.) with its LED connected to the collector of Q1 through a resistor. In this configuration the R3 resistor and LED must be connected in series and the value of R3 must be recalculated so, that an optotriac's LED will be turned on when Q1 is open. Naturally, you will not need the stage with the POWER MOSFET Q2 anymore. You may also substitute for the LM567C chip using LM567 or NE567.

Below (Fig.4.6) you can see my assembled circuit. As an external load I used a PC's 12V fan connected to the drain of the POWER MOSFET IRF510. Also I took a control signal from the single channel of a sound card (you may apply both audio channels if driving two loads at a time, but this also requires using two tone–decoders).

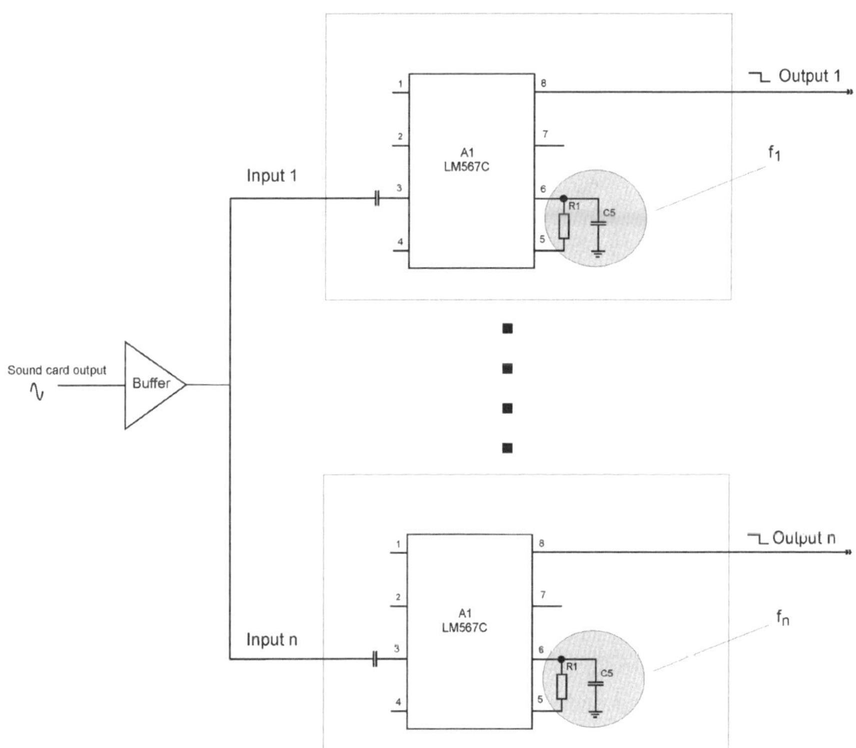

Fig.4.6

Using the above basic circuit you can drive several channels simultaneously. In this case you should introduce several modules; each with a LM567C chip, Each of them will control its own load. The inner oscillator of each module has to be configured for operating with its own unique frequency by selecting the appropriate values of your R1–C5 pairs (see Fig.4.5). The functional diagram of such a multichannel control system is shown in Fig.4.7:

Fig.4.7

In this circuit the sound card produces various frequencies which are delivered through a buffer to the n inputs of n tone-decoder modules. Each of those n modules is configured to sense a particular frequency f_1 ... f_n. A sound card in this configuration must generate a pure sinusoid signal whose frequency may be one of f_1 . . f_n. For reliable functioning the pass-bands of the tone–decoders must not overlap – in this case the low–level active signal appears only on one output at a time for each predetermined input frequency. As a sound card is capable of generating signals in the range of 20 – 20000 Hz, you can thus use multiple modules for driving multiple loads.

The audio signal generated by a sound card may be easily converted into a DC voltage. We can build circuits where this voltage will change linearly when the input frequency varies over some range. Generating a DC voltage controlled by the frequency may be of use when designing robotic systems, for testing various circuits and motor handling, etc. To achieve a high precision use a frequency–to–voltage (F-to-V) converter with high linearity. One low-cost and effective solution is based on the popular chip LM131/231/331. This part may implement both V–to–F and F–to–V conversion providing high linearity of about 0.01%. The schematic circuit of such a F-to-V converter is shown in Fig.4.8:

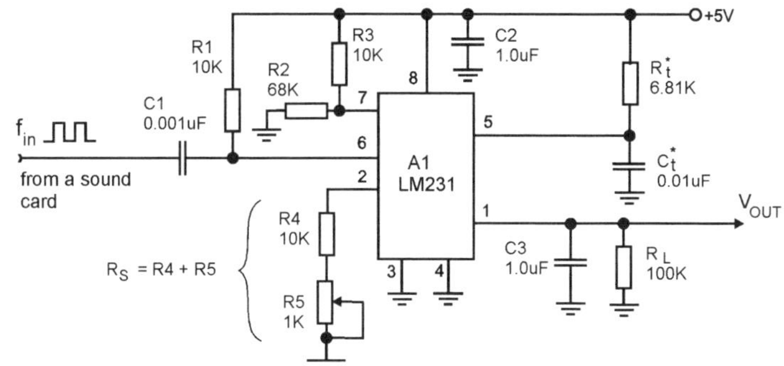

$$V_{OUT} = f_{in} \times 2.09V \times \frac{R_L}{R_S} \times R_tC_t$$

Fig.4.8

This circuit functions quite well up to a frequency of 3 KHz. The frequency range is determined by the Rt– Ct pair; for other ranges you should recalculate values of Rt and Ct using the formula in Fig.4.8.

The C3 and R_L pair represents the simple low-pass filter suppressing the noise riding on a useful signal V_{OUT} at the output of the LM231 (pin 1). The input frequency f_{in} should be the square wave, so you must previously convert the sound card's output signal into square pulses. A higher precision output voltage V_{OUT} may be reached when applying an active low-pass filter instead of the C3–R_L pair. You may also apply other F–to–V chips when building similar circuits. For example, it's possible to build the circuit using the low-cost AD537 chip.

Let's see how to operate with sound and build embedded projects using MATLAB. In many aspects MATLAB simplifies the design of those projects, for it provides an easy program interface for writing control software.

Processing low-level input signals

Sometimes you may wish to capture low-level audio signals generated by various transducers that measure some physical quantity (light, proximity, vibration, etc.). We can certainly pick up those signals using a microphone or line input of a sound card, provided that we place in an additional gain stage between the output of the transducer and the input of the sound card. This allows amplifying of the weak signals before they go to an internal preamplifier of a sound card. This preamplifier usually has a limited sensibility (a low signal-to-noise ratio, SNR), so an additional external amplifier comes in handy. For instance, when dealing with a vibration sensor I applied a simple amplifier with a low-cost and widely used TL081 opamp (Fig.4.9):

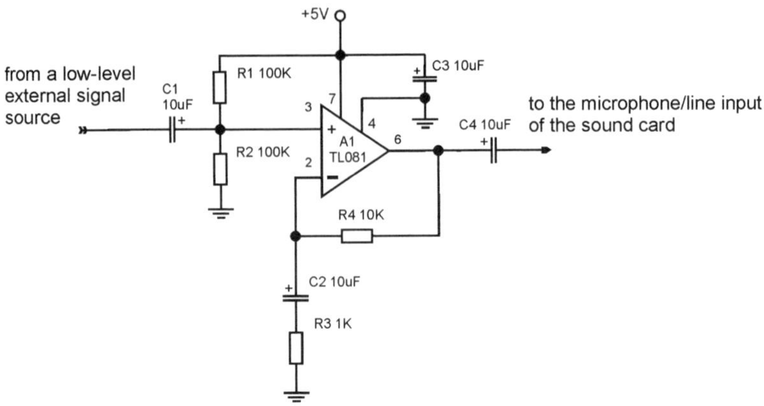

Fig.4.9

The opamp is connected as a non-inverting amplifier for AC signals – the R3, R4 and C2 components form the feedback circuit. The capacitor C2 connected in series with R3 exhibits very low impedance for audio signals, so the gain for AC signals may be calculated as:

```
Gain = 1 + R4/R3
```

In our case, the gain is 11V/V, however, you can easily adjust it by changing values of R3 and R4. It would be wise to take R3 and R4 in the range several KOhm through to several tens of KOhm. Remember that low values of resistors result in a greater power consumption, while higher resistances increase overall noise of this circuit.

For DC signals the opamp is connected as a voltage follower. The voltage divider (R1-R2) provides a bias to the non-inverting input of A1 equal to half of the supply voltage (2.5V). The signal from the source is fed to the non-inverting input through C1; the amplified signal goes to the sound card input via capacitor C4.

Processing sound in MATLAB

Operating with audio signals in MATLAB is extremely simple for it provides an excellent programming interface to process audio data. There are lots of functions which allow manipulating of audio data. They are described in detail in the appropriate reference manual. But let's see some practical examples.

To read the microphone input and then play this via the line-out output of the sound card you may use the following sequence (Listing 4.5):

Listing 4.5

```
>> ro = audiorecorder(11025, 16, 1);
>> record(ro, 5);
>> pause(ro);
>> p = play(ro);
```

Let's examine this code. First we create an audio recorder object through which we can operate with audio data. The statement

```
>> ro = audiorecorder(11025, 16, 1);
```

creates the `audiorecorder` object `ro` using the sampling rate 11025 Hz, the sample size 16 bits, and the number of channels equal to 1 (mono mode). The `audiorecorder` function could be invoked without parameters as is shown below:

```
ro = audiorecorder;
```

When the audiorecorder function is called without parameters it is assumed that this function takes default arguments: sampling rate 8000 Hz, sample size 8 bit, and 1–channel for sound processing. The default parameters are useful when operating with most sound cards, because this configuration is supported by just about all devices.
As to the sampling rates, they can take the following values: 8000, 11025, 22050, and 44100. The sample size must be 8 or 16 if we deal with Linux OS and the number of channels may be 1 for mono mode or 2 for stereo mode. The MATLAB documentation hints that the current implementation of `audiorecorder` is not intended for long, high–sample–rate recording because it uses system memory for storage and does not use disk buffering. When large recordings are attempted, MATLAB performance may degrade. There are various methods (mainly described at MATLAB forums) for improving performance, so I encourage you to investigate those approaches.
Let's back to our code. After the `audiorecorder` object `ro` has been created we can begin recording sound during the next 5 seconds from the microphone/line input (you should set the input source before using). The following statement does this:

```
>> record(ro, 5);
```

After pausing we play back the sound just recorded by invoking a `play` function:

```
>> p = play(ro);
```

As you can see the recording / playing of sound is not complicated, because MATLAB executes all the operations for you using just a few functions. More complicated algorithms may be realized when we involve timers and program loops.
The following example exhibits how to generate a sound waveform programmatically. We already considered how to create a sinusoid waveform in C (see Listing 4.1). Implementing the same task in MATLAB is extremely simple (Listing 4.6):

Listing 4.6

```
>> cf = 1000;
>> sf = 11025;
>> d = 3.0;
>> n = d * sf;
>> s = (1:n)/sf;
>> s = sin (2 * pi * cf * s);
>> sound (s, sf);
```

Here the `cf` variable holds the frequency in Hz (1000Hz, in our given case). The `sf` variable determines a sampling rate equal to 11025 samples per second. Besides the variables above we need to set a time interval (duration) for sounding. The sound will last 3 seconds, so we save 3 in the `d` variable. The following statement allows saving a total number of samples needed for playing in the `n` variable. This value is determined as a product of `d` and `n`. To understand this, you may recall the curve in Fig.4.2.
The next step is to create a buffer where the samples will be held. Obviously the size of such a buffer must be at least equal to `n` to hold all possible values. The buffer is defined as an array `s` which values runs from `1/sf` to `n/sf`:

```
>> s = (1:n)/sf;
```

As we want to play a sinusoid waveform we should design it as is shown below:

```
>> s = sin (2 * pi * cf * s);
```

Finally, we can play a sound buffer using the `sound` function:

```
>> sound (s, sf);
```

Besides sinusoids you can create other waveforms (square, triangle, sawtooth, etc.), although this involves a bit more maths than in the simple case of the pure sinusoid.

When we want to generate a continuous sound in MATLAB we may apply a timer object. The following example demonstrates this approach. First of all let's define the callback function (name it as playSnd) for our timer and save the code of this function in a function M-file that will be named as playSnd.m. The source code of the playSnd function you can see below in Listing 4.7:

Listing 4.7

```
function playSnd(obj, event)
%UNTITLED Summary of this function goes here
%   Detailed explanation goes here
cf = 1000;
sf = 44100;
d = 1.0;
n = sf * d;
s = (1:n) / sf;
s = sin(2 * pi * cf * s);
soundsc(s, sf);
end
```

The playSound function will be invoked each time the timer fires. The second part of our software is located inside M-file and has the following source code (Listing 4.8):

Listing 4.8

```
t = timer('StartDelay', 5, 'Period', 5, 'TasksToExecute', 5, 'Ex-
ecutionMode', 'fixedRate');
t.StartFcn = 'disp(''Timer started'')';
t.StopFcn = 'disp(''Timer stopped'')';
t.TimerFcn = 'playSnd';
start(t);
```

The very first statement creates the timer object t. After an initial delay of 5 s the timer starts firing every 5 s (the Period parameter) until 5 such cycles have passed (the TasksToExecute parameter). To complete the configuration of the timer t let's set its mode to 'fixedRate'. The StartFcn and StopFcn functions allow display of appropriate messages after the timer t has started or stopped, respectively.

The TimerFcn parameter is assigned a 'playSnd' which corresponds to the playSnd callback function saved in playSnd.m file. As you know from the previous chapter TimerFcn determines a function that will be executed when a timer fires. That's all about configuring the timer object t.

The statement

```
start(t)
```

starts our timer running.

By setting the `TasksToExecute` parameter to an appropriate value you can control the interval when a timer is active.
We can also convert a given waveform into square or sawtooth by using the circuit shown in Fig.4.4. Thus we can have three types of signals (sinusoid, square and sawtooth) at a time using a single channel of a given sound card.
The last topic we will discuss in this chapter relates to visualization and estimation of audio data. The simplest way to see and estimate an audio signal is to take an oscilloscope, place a probe at the input or output of the sound card and capture the signal. However, the oscilloscope is an expensive device; so many hobbyists don't have one. Fortunately, there are several software methods of viewing and measuring the parameters of a sound signal. As to Linux, in order to observe an audio signal we can apply one of the free programs usually known as "audio" or "sound" editors.

Here we should mention two approaches to the visualization and estimation of audio signals. First, when we try to observe a signal arriving at the input of the sound card, we simply switch to the "record" mode usually available in all audio editors and then capture the audio signal continuously or during some predetermined time. Then we can save the captured signal to some WAV-file and/or start processing audio data. This is a common way of dealing with the input audio data when using Audacity, GoldWave or many other popular audio editors.

Another approach may be taken when we want to estimate a signal that our sound card yields on its output. We can connect the output to the input ("line-in") using an audio cable, then capture the audio signal generated by a sound card and evaluate it. The sound may be created programmatically using a C++ program or launching, for example, Matlab and then may be captured by the audio editor, such as Audacity or GoldWave. When you apply this method, you have to configure your sound card to operate in a duplex mode – in that case the built-in Digital-To-Analog (DAC) converter of the sound card will generate a data audio stream that will be picked up by the Analog-To-Digital (ADC) converter responsible for digitizing the data. Obviously, the input and output audio data must have the same format (sampling rate and number of bits per sample) to avoid errors. It is also desired that the power levels of both input and output signals should be set to a mid-scale.
Let's see in the following examples where we apply a Matlab environment to write a program to generate sound. Remember that the output of the sound card must be linked to the "line-in" input. In these experiments I used an "old" ESS1869 sound card plugged into an "ancient" Pentium 4 PC; this configuration, however, exhibited satisfactory performance. The following source code (Listing 4.9) saved in a `pure_sinwave_test.m` file generates a sine waveform:

Listing 4.9

```
cf = 500;
sf = 8000;
d = 10.0;
n = sf * d;
s = (1:n) / sf;
s = sin(cf * s);
soundsc(s, sf, 16);
```

We configured our script so that the sampling rate of the signal generated is equal to 8000 (an `sf` variable) and the number of bits per sample is 16. The duration is selected to be equal to 10 seconds. Before executing this file we must launch the Audacity audio editor and configure it to capture an audio signal with a sampling rate of 8000 and 16 bits per sample. After all the preparations are in place we can execute our m-file:

```
>> pure_sinwave_test
```

Simultaneously, while launching the above script we start recording the sound. After the operation has completed we can see the data obtained with Audacity (Fig.4.10):

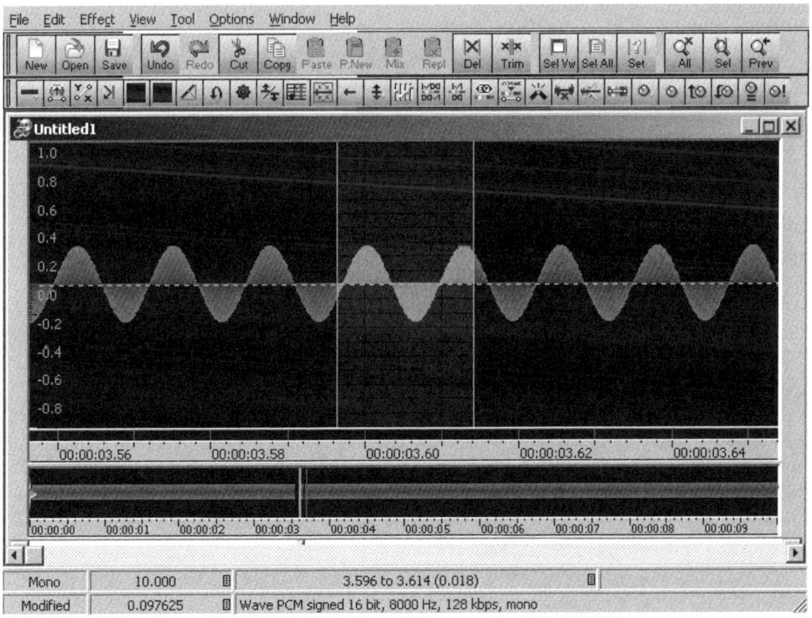

Fig.4.10

We can also take another very popular sound editor called GoldWave (a trial version is available). The result of capturing the same audio signal in GoldWave will be as shown in Fig.4.11:

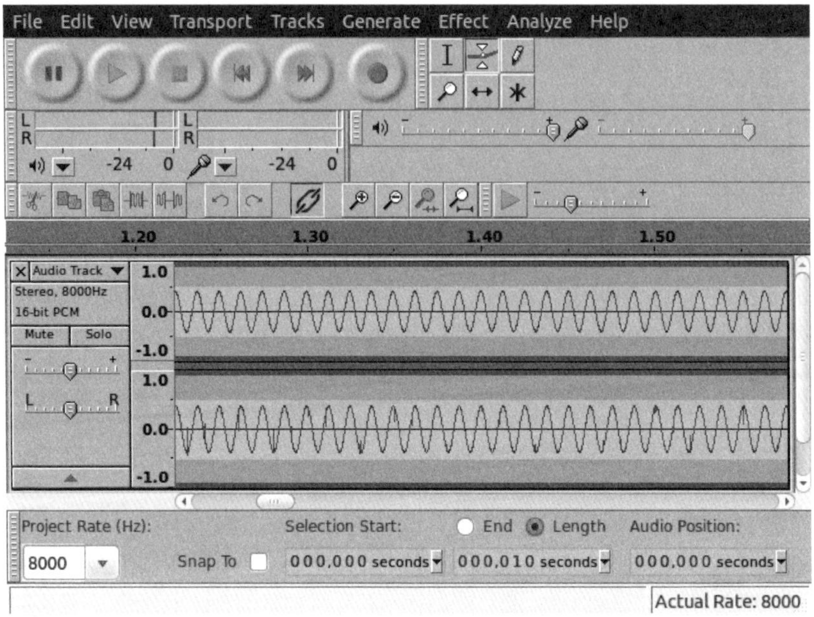

Fig.4.11

In the second example our sound card generates the signal that resembles a square wave by applying a Fourier series method. Any periodic function can be expressed as a series involving sines and cosines, known as a Fourier series. The square wave signal may be represented as a sum of variously weighted odd harmonics of the sine function. By using this approach we can represent, for example, the square wave as shown below:

```
Square wave  = 4/π * [(sin(x)/1 + sin(3x)/3 + sin(5x)/5 + sin(7x)/7
+ ...]
```

From a practical point of view we can take the first four terms from the above expression to design an approximate square wave. Below you can see a full listing of the program in Matlab (Listing 4.10):

Listing 4.10

```
cf = 500;
sf = 22050;
d = 10.0;
n = sf * d;
s = (1:n) / sf;
s = 1.3 * sin(cf * s) + 0.45 * sin(3 * cf * s) + 0.25 * sin(5 * cf
* s) + 0.18 * sin(7 * cf * s);
soundsc(s, sf, 16);
```

We save this source in a `square_wave_test.m` file and execute this script together with any audio editor. In the case of using Audacity this gives us the following picture (Fig.4.12):

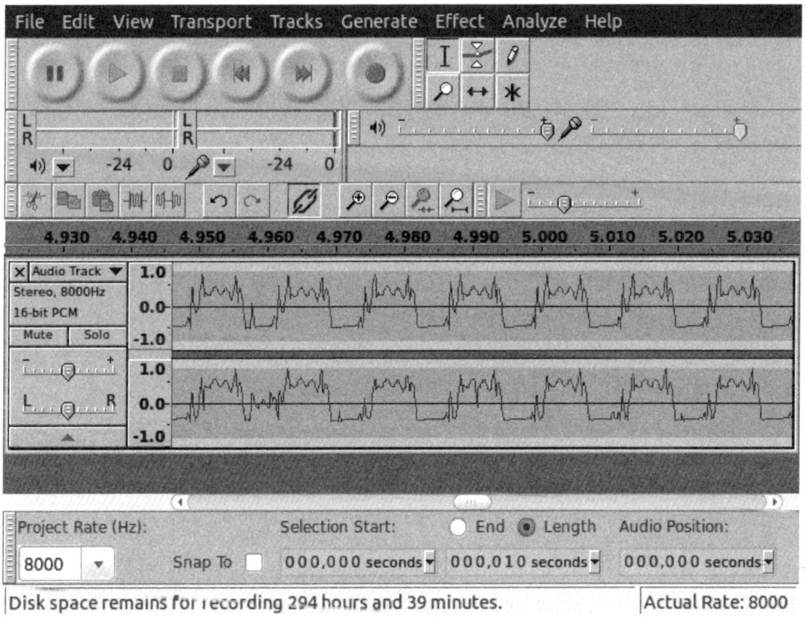

Fig.4.12

After you have configured Audacity you can launch the m-file to execute:

```
>> square_wave_test
```

Launching the GoldWave application with the above program gives the following result (Fig.4.13):

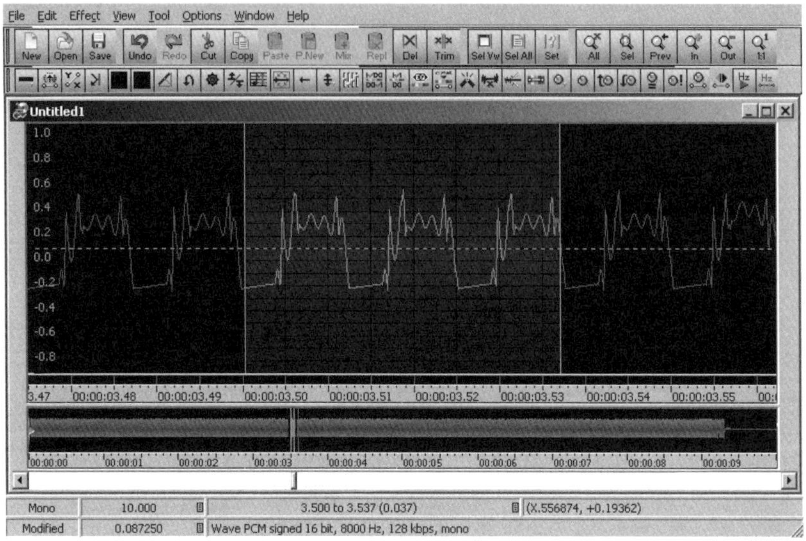

Fig.4.13

If you introduce, for example, 9th and 11th harmonics from the expression, thus,

```
Square wave  = 4/π * [(sin(x)/1 + sin(3x)/3 + sin(5x)/5 + sin(7x)/7
+ ...]
```

into an appropriate place in Listing 4.50 (where the signal `s` is determined), you will gain a better approximation for our square wave signal.

By now we already know much about audio data processing. The only question left is how to save our sounds onto the disk. There are various formats of files to store audio data. You may write your audio buffer into a binary file (called "raw" file), or may want to store data in a special format (.wav, mp3, etc.) used to order the audio data in some way. The advantage of "pure" raw audio files is that they can be easily converted into any specific format by using audio editor applications. In order to show some basic operations with sound files a very simple application in C++ was developed.

Its source code is shown below in Listing 4.11. The program picks up the audio data at the sound card input (a “Line-in” jack) and writes data to a .raw file. Here it is assumed that the default audio system used is ALSA, so we can apply ALSA API functions to process the sound signals. Let’s title our audio file as a Sound_Test_22050.raw (generally, we can omit the .raw extension when writing to a binary file). The sampling rate is supposing to be 22050.

Listing 4.11

```
#include <stdio.h>
#include <unistd.h>
#include <stdlib.h>
#include "alsa/asoundlib.h"

int main (void)
{
 int i;
 int err;
 short buf[32000];

 unsigned int rate = 22050;
 snd_pcm_t *capture_handle;
 snd_pcm_hw_params_t *hw_params;

 FILE* fp;
 fp = fopen("Sound_Test_22050.raw", "ab+");

 if ((err = snd_pcm_open (&capture_handle,
                    "dcfault",
                     SND_PCM_STREAM_CAPTURE, 0)) < 0)
 {
  fprintf (stderr, "cannot open audio device: %s\n",
             snd_strerror (err));
  exit (1);
 }
 if ((err = snd_pcm_hw_params_malloc (&hw_params)) < 0)
 {
  fprintf (stderr, "cannot allocate hardware
          parameter structure: %s\n", snd_strerror (err));
  exit (1);
 }
```

```
 if ((err = snd_pcm_hw_params_any (capture_handle,
                                   hw_params)) < 0)
  {
   fprintf (stderr, "cannot initialize hardware
              parameter structure: %s\n", snd_strerror (err));
   exit (1);
  }

if ((err = snd_pcm_hw_params_set_access (capture_handle,
                                         hw_params,
            SND_PCM_ACCESS_RW_INTERLEAVED)) < 0)
{
 fprintf (stderr, "cannot set access type: %s\n",
             snd_strerror (err));
 exit(1);
}
if ((err = snd_pcm_hw_params_set_format (capture_handle,
                              hw_params,
                              SND_PCM_FORMAT_S16_LE)) < 0)
{
 fprintf (stderr, "cannot set sample format: %s\n",
            snd_strerror (err));
 exit(1);
}

if ((err = snd_pcm_hw_params_set_rate_near (capture_handle,
                                        hw_params,
                                        &rate, 0)) < 0)
{
fprintf (stderr, "cannot set sample rate: %s\n",
            snd_strerror (err));
exit (1);
}
if ((err = snd_pcm_hw_params_set_channels (capture_handle,
                                    hw_params, 1)) < 0)
{
```

```
fprintf (stderr, "cannot set channel count: %s\n",
              snd_strerror (err));
exit(1);
}

if ((err = snd_pcm_hw_params (capture_handle,
                         hw_params)) < 0)
{
 fprintf (stderr, "cannot set parameters: %s\n",
              snd_strerror (err));
  exit(1);
}
        snd_pcm_hw_params_free (hw_params);
if ((err = snd_pcm_prepare (capture_handle)) < 0)
{
fprintf (stderr, "cannot prepare audio interface
              for use: %s\n", snd_strerror (err));
exit(1);
}

for (i = 0; i < 200; i++)
{
 if ((err = snd_pcm_readi (capture_handle,
                          buf,
                          32000))!= 32000)
{
  fprintf (stderr, "read from audio interface failed: %s\n",
                 snd_strerror (err));
  exit(1);
}
fwrite(buf, sizeof(short), 32000,fp);
}
snd_pcm_close (capture_handle);
fclose(fp);
return 0;
}
```

Let's talk about variables first. The buf variable identifies the array of 32000 bytes where the audio data is stored and the rate variable determines the sampling rate. The `capture _ handle` variable will contain the handle to the audio device currently used and the `hw _ params` variable points to the structure where all parameters of the data of the audio device are stored.

The preparation steps before capturing audio data are implemented in a common way. First,the program opens the audio device set as default in the Linux system (the `snd _ pcm _ open` function), then appropriate parameters are written into `hw _ params`. The audio device is configured to operate in the mono mode with the sampling rate 22050, 16 bits per sample. After writing the parameters to the sound card (the `snd _ pcm _ hw _ params` function) and preparing an audio interface for data reading (the `snd _ pcm _ prepare` function) the program starts capturing audio data within a `for()` loop with a counter of 200.

At the beginning of each iteration the data stream is read into `buf` (the `snd _ pcm _ readi` function), then the content of buf is stored into a file stream `fp` (the `fwrite()` function) previously opened for a binary writing by a function `fopen()`:

```
fp = fopen("Sound_Test_22050.raw", "ab+");
```

After the loop has ended the audio device descriptor is closed (the `snd _ pcm _ close` function). In order to flash data to the disk we need to close the file pointer `fp` (the `fclose()` function) as well. The .raw file just created can be played back using an `aplay` standard utility included in the ALSA system. You may also apply the popular sox software interface for reproducing this file. In addition, the .raw file can be played back by any popular program such as GoldWave as well. Note, that the audio data recorded through the line-in input may lack high quality because of low performance characteristics of a preamplifier stage of the input, so we usually need to get rid of the noise by applying some advanced audio editor such as GoldWave, Audacity, etc. You may experiment with this source code and improve it if necessary.

When recording/playing the audio data we can also use two basic utilities provided by the ALSA system. These are `arecord` and `aplay`. Both are console applications and have a lot of parameters described in the relevant man-pages. Being command line programs, both utilities may be easily put into shell scripts. The `arecord` program is a command-line sound recorder for the ALSA sound card driver. It supports several file formats and multiple sound cards with multiple devices. The aplay utility is much the same, only it plays audio data instead of recording.

For supported sound file formats, the sampling rate, bits per sample, and so forth can be automatically determined from the sound file header. Let's see a brief example on how to apply those utilities in practice. This example shows the use of both utilities for capturing and recording popular music on the Internet. We will capture two audio streams at www.youtube.com/Spocktrometer and save them in WAV-files. I took two things, "Pump" and "House full of bullets" and convert the audio stream of both into two WAV-files using the `arecord` utility.

In order to capture the sound stream we can apply either the program mixer that allows passing the audio data from the Analog-To-Digital converter (ADC) of a sound card straight to the Digital-To-Analog converter (DAC) responsible for capturing sound. This will work, however, only for sound cards with a "loopback" interface where the audio data stream may be redirected from the internal ADC to DAC. Alternatively, we may choose another way – we will connect the "Speaker out" output and the "Line-In" input by using an audio cable. In that case the sound that emerges at the output goes straight to the line-in input.
When the audio stream from the Internet reaches the speaker output, we will capture the sound and record it into a WAV-file using the `arecord` utility. Again, as the input preamplifier stage of the sound card is noisy, we should wipe the noise within a WAV-file by launching some audio editor. For capturing and recording music, the `arecord` utility has to be executed with the following parameters:

```
$ arecord -d 400 -c 2 -f dat -t wav The_Pump.wav
Recording WAVE 'The_Pump.wav' : Signed 16 bit Little Endian,
Rate 48000 Hz,Stereo
```

After recording is complete the current directory will contain the The_Pump.wav file. In the same way the following music file is written:

```
$ arecord -d 400 -c 2 -f dat -t wav The_House_full_of_bullets.wav
Recording WAVE 'The_House_full_of_bullets.wav' : Signed 16 bit
Little Endian, Rate 48000 Hz, Stereo
```

To play a WAV-file just saved we can apply the `aplay` command:

```
$ aplay -c 2 -f dat -t wav The_Pump.wav
Playing WAVE 'The_Pump.wav' : Signed 16 bit Little Endian,
Rate 48000 Hz,Stereo
```

Note that the transfer of audio streams through the Internet requires a high network performance. Another way to record/play sound is to apply the sox utility. Sox reads and writes audio files in most popular formats and can optionally apply effects to them; it can combine multiple input sources, synthesize audio, and, on many systems, act as a general purpose audio player or a multi-track audio recorder. It also has limited ability to split the input in to multiple output files. Finally, we can professionally process audio signals when applying some of popular audio editors with the graphical user interface (Audaicity, GoldWave and so on.)

Conclusion

We discussed some uses of the standard sound card in Linux operating systems. As you can see, this popular device may be fitted in the control or measurement system. The sound card may successfully control the external circuitry, may serve as the generator of low frequency signals or may process lowlevel input signals. By combining the sound card with other peripherals such as a serial or parallel port of the PC, you may design powerful data acquisition and measurement systems. The sound card is well supported by popular software packages such as NI LabView and Matlab/Simulink thus allowing us to develop complex projects.

Chapter 5. Network and wireless systems for home and laboratory

So far we discussed designing miscellaneous embedded projects working locally on your desktop or laptop. But often nowadays you also may wish to control your hardware or get data from some remote place. This may be required when, for example, you want to operate your system at home from your workplace or vice versa.. In these cases you need to build some distributed system consisting of some relatively independent parts (subsystems) where each of them may function on a standalone PC. You might also wish to modify an embedded project already built so that it would be possible to pass data to another application, such as MySQL, OpenOffice or a text editor running on the same machine or somewhere on the Internet. Putting your local application in some distributed environment may provide new facilities for your system. For example, you may arrange your distributed system so that one application would execute some low-level data processing while another application would accomplish high-level operations such as visualization of results and/or providing a nice graphical user interface. The above examples illustrate just a few numerous situations where you start thinking about building a distributed system. This chapter will help you in understanding many aspects of designing such distributed systems in Linux.

In general, there are some common methods for developing distributed systems in Linux – we will discuss two of them at the beginning of this chapter with the remainder dedicated to the practical aspects of designing various distributed systems. The methods herein are based on using networking features of Linux. These common features are TCP/IP and local sockets that underlie networking.

Now, when referring to the "distributed application" we will mean the "network application" and vice-versa.–. Making distributed systems requires a lot of network programming and knowledge of, at least, the basic concepts of development of network applications. The key term in network programming is "socket", so we will consider it in detail before detailing our application.
The term "socket" may be regarded as an "address" or "endpoint" assigned to a network application. The TCP/IP socket is determined by two basic parameters, IP-address and port. Two network applications may interact with each other only via sockets.

The IP-address identifies a PC on the network and should be unique for the particular network. The port is a number which when coupled with the IP-address allows identifying the socket for the particular network application. This is a very simplified description, but in spite of this it should help make things clearer. An interaction between two network applications may be illustrated by Fig.5.1.

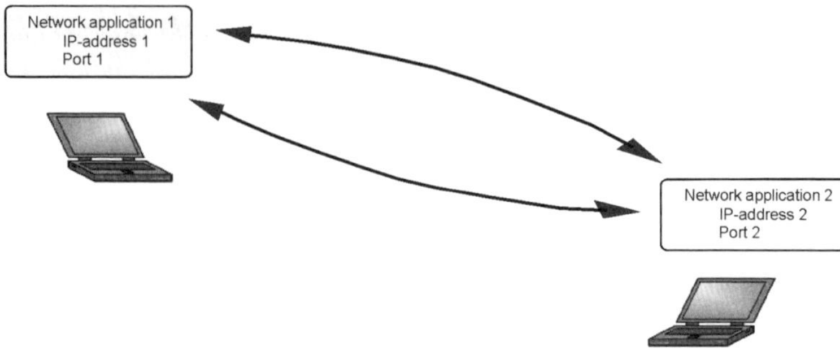

Fig.5.1

You can see how "Network Application 1" communicates with "Network Application 2" over the network. The "Network Application 1" wants to talk to the "Network Application 2" and therefore it must first know the IP-address 2 and Port 2. Otherwise, "Network Application 2" when dealing with "Network Application 1" should know about IP-address 1 and Port 1. The IP-address of the network application is the same as that of the network interface of the PC where the application executes.

Let's discuss IP-addresses and ports in detail.
An IP address is an identifier that is assigned to the "network interface" or a set of interfaces. The "network interface", from the first point of view, appears to the user as the network card of your PC together with appropriate software that controls data transfer via this device. Each IP-address can identify the source or destination of IP packets (data blocks travelling over the network). Nowadays many IP-addresses are formed according to the IPv4 standard, though; in the future ipv4 will be replaced with IPv6. The reason for this is obvious: IPv4 addresses are just 32 bits long but an IPv6 address is 128 bits long, which is four times larger than an IPv4 address. This allows us to significantly extend the Internet address space that otherwise is about to be overflowed.

When enabling TCP/IP on a network interface we should assign to it one or more logical IPv4 addresses, either automatically or manually. The IPv4 addresses may be represented as 32-digit strings consisting of 0s and 1s, but this is so cumbersome, that dotted decimal notation was introduced, in which periods (dots) separate four decimal numbers from 0 to 255. Each decimal number (octet) represents 8 bits or 1 byte of the 32-bit address. For example, the IPv4 address 1100001000001111 0001110000001010 is expressed as 194.15.28.10 in a dotted decimal notation.

Besides the IP–address the application must supply the TCP port number of the destination application. The port provides a location for sending segments. Each port must be identified by its unique number. Port numbers below 1024 are the well known ones assigned by the Internet Assigned Numbers Authority (IANA) for predefined network applications. For instance, a port with number 23 is reserved for Telnet; port 80 is reserved for HTTP, etc.

By now we know how network applications communicate each with other. The next question is how to program a network application to transfer or receive some data. To make things clear, it will be useful to see the hierarchy of network applications. Every day you are faced with various types of network applications. A widely used web browser is an example of such a kind of application.. As a general rule for programming network applications, a special four-layer conceptual model that is known as DARPA model was introduced.
This model maps the TCP/IP protocol suite and includes Application, Transport, Internet and Network Interface layers (Fig.5.2):

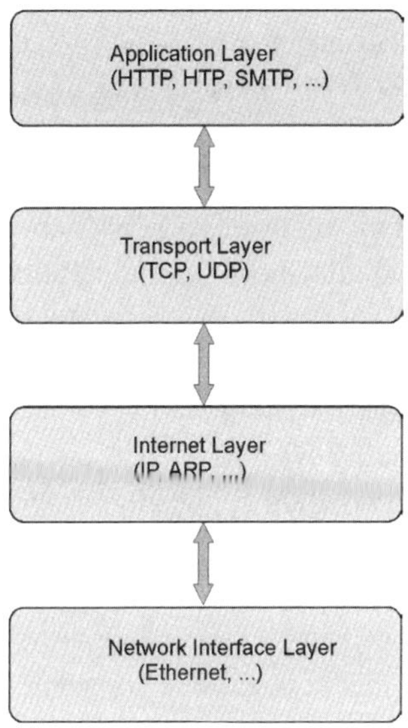

Fig.5.2

This model allows an understanding of how the network applications interact with each other.
The Network Interface layer (also called Network Access layer) provides for sending TCP/IP packets on the network medium and receiving TCP/IP packets off of the network medium. Note, that the TCP/IP suite was designed to be independent of the network access method, frame format, and medium. Therefore, we can use TCP/IP to communicate across differing network types that use various LAN technologies (Ethernet, 802.11 wireless LAN, WAN technologies). Because of its independence from any specific network technology TCP/IP can be also easily adapted to new technologies.

The Internet layer is responsible for addressing, packaging, and routing. Amongst other things the Internet layer includes the Internet Protocol (IP) that addresses, routes, fragments, and reassembles packets.
Next, a Transport layer provides the Application layer with session and datagram communication services. The core protocols of the Transport layer are TCP and UDP. TCP provides a one-to-one,

connection-oriented, reliable communications service. TCP establishes connections, sequences and acknowledges packets sent, and recovers packets lost during transmission. In contrast to TCP, UDP provides a one-to-one or one-to-many, connection-less, unreliable communications service. UDP is used when the amount of data to be transferred is small (for example, a data block that would fit into a single packet), when an application developer does not want the overhead associated with TCP connections, or when the applications or upper-layer protocols provide reliable delivery. TCP and UDP operate over both IPv4 and IPv6 Internet layers. In our projects we will apply TCP/IP only.

The Application layer allows applications to access the services of the other layers, and it defines the protocols that applications use to exchange data. The Application layer contains many protocols, and more are always being developed. The most widely known Application layer protocols help users exchange information via the Internet. The well known Hypertext Transfer Protocol (HTTP) transfers files that make up pages on the World Wide Web. The File Transfer Protocol (FTP) transfers individual files, typically for an interactive user session and the Simple Mail Transfer Protocol (SMTP) transfers mail messages and attachments. Additionally, some Application layer protocols help to manage TCP/IP networks.

How, in practice, do those layers relate to writing our LAN applications? To answer this question we should consider how Windows operating systems support the development of network applications. Windows network applications use two main application programming interfaces (APIs) to access TCP/IP services: Windows Sockets and NetBIOS. Nowadays the Windows Sockets is the most commonly used modern API for developing network applications in Windows. Almost all TCP/IP applications such as services and tools supplied with Windows are examples of Windows Sockets applications. Windows Sockets provides services that allow applications to use a specific IP-address and port to initiate and accept a connection to a specific destination IP-address and port, send and receive data, and close a connection.

There are three types of sockets that may be used in network applications:

- stream sockets, which provide a two-way, reliable, sequenced, and unduplicated flow of data using TCP.
- datagram sockets, which provide bidirectional flow of data using UDP.
- raw sockets, which allow protocols to access the IP packet directly, without using TCP or UDP.

As mentioned earlier, the socket functions are the endpoint for network communication. Recall that two sockets, one for each end of the connection, create a bidirectional communication path.
An application creates a stream or datagram socket by specifying the IP–address of the host and the port the application is using. When developing the network application you also need to specify the type of service (TCP for connection-based service and UDP for connectionless). Once we are going to use TCP protocol in our applications, we need to specify that explicitly.
The TCP protocol has a great advantage compared to other protocols, such as UDP, IP, etc. because it provides for correct delivery of data from the PC to the PC. Because data can be lost somewhere

in the network, TCP adds support to detect errors or lost data and to trigger re-transmission until the data is correctly and completely received.
The internet offers many sources of documents which introduce network communications with TCP protocol stacks if you are need detailed reading into the subject. We will end network basics here and discuss one more concept that needs understanding before writing network applications. This concept describes the "client–server" architecture that is common for network communications. The "client–server" architecture is commonly used in designing distributed systems.

Client–Server architecture

When running, the network application may function as a “server” or as a “client”. Many times the network application is developed for operating in both modes, depending on some conditions. An example of the “client–server” configuration is shown in Fig.5.3:

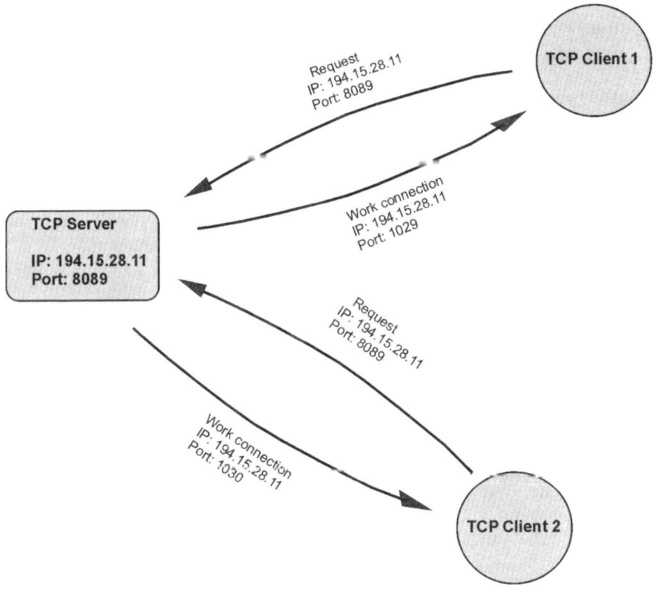

Fig.5.3

Here you can see an interaction between three network applications (one TCP server and two TCP clients). The TCP server has the IP-address 194.15.28.11 and works with port 8089 where it waits (“listens to”) for incoming requests from TCP clients.

Suppose that the TCP Client 1 located somewhere on the network tries to connect to the TCP Server. Initially, the Client 1 asks the IP-address 194.15.28.11 and port 8089 belonging to the server. The TCP Server discovers that request and creates a working connection (working socket) to the TCP Client 1. Besides, the server specifies the parameters of this connection (IP-address and port) and passes them to the client. As you can see, the working socket has the IP-address of the TCP Server, but a different port (1029, in our case). Data transfer between the TCP server and the TCP Client 1 will be accomplished via this working socket which must be closed after operations have completed.

Again, when the TCP Client 2 establishes connection to the TCP Server, the new working connection with IP-address 194.15.28.11 and port 1030 is opened. Note, that the "listening socket" working on port 8089 is not involved in the data transfer process between the TCP Server and clients. This socket only spies for incoming requests from clients and forces the TCP Server to create a new working connection when needed. When the TCP Server completes its functions it must close the "listening" socket as well.
Usually, we need at least two PCs (one server and one client) to realize client–server configuration.

However, we can build and test our client-server configuration using a single PC. This is because all operating systems reserve a so-called "loopback interface" with the IP-address 127.0.0.1. This interface is frequently used when testing network applications or when debugging the TCP/IP configuration. This IP-address doesn't tie to any network card; it is a pure "virtual" interface. We will develop our distributed systems described in this chapter using this "loopback interface".
Note that in real life you must use real IP-addresses, so you should replace "127.0.0.1" with a valid address in the source code of your applications or use names of hosts if you have some name resolution system.

Many times you may want to have a distributed system functioning locally on your PC. For example, this may be the case when some application provides measurement of some physical quantities (temperature, pressure, light, humidity, etc.) and passes data to another application which executes high-level processing (e.g. OpenOffice, MySQL, etc.). These systems may interact with each other via TCP sockets, but there is actually another method based on UNIX domain sockets known also as IPC (inter-process communication) sockets. Often those sockets are called "local sockets".

A local socket is a data communications endpoint that is similar to an Internet (TCP) socket, but does not use a network protocol for communication. Local sockets were introduced in POSIX operating systems, including Linux, for inter–process communication.
UNIX domain connections appear as byte streams, much like network connections, but all data remains within the local computer. UNIX domain sockets utilize the file system as address name space, i.e. they are referenced by processes as *inodes* in the file system. In this way two standalone processes may open the same local socket in order to communicate with each other. In real life, the actual communication where data travels across local sockets does not use the file system, but buffers in kernel memory.

The simplified diagram of communication of two processes via local sockets is illustrated by Fig.5.4:

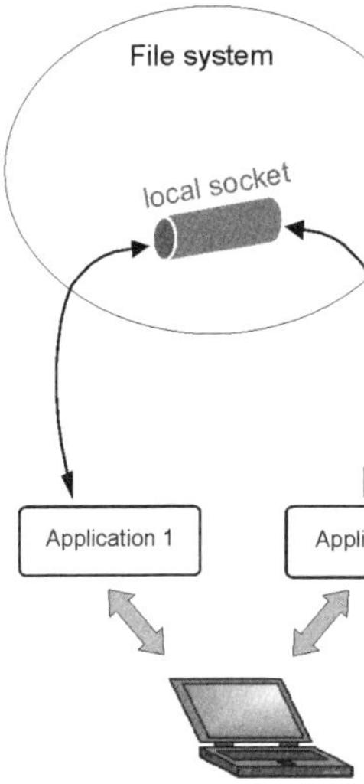

Fig.5.4

UNIX domain sockets consume much less system resources compared with TCP/IP or UDP sockets because they don't need to calculate checksums, insert headers, perform routing, etc as in the case of using TCP/IP protocol.
In general, if you are looking for performance locally, UNIX domain sockets probably will be the best choice. Using local sockets helps standalone applications written in various languages to communicate with each other within the same PC. Local sockets are simpler in programming than TCP sockets.

Simple client–server system

The Client–Server approach may be successfully applied while developing remote or distributed embedded systems. To become more familiar with this concept let's discuss some practice examples where we will program client and server applications.
Before dealing with network applications we need to determine the IP–addresses and port so that both client and server will interact with each other. As we are going to experiment with our network systems using the same PC, the IP–address for client and server will be 127.0.0.1 which is the address of a loopback interface. As to the port number, we can take the first one available in the Linux operating system. In my system I took port 7171.

This port will be used for "listening" by our TCP server application. Both IP–address and port appear as an "endpoint" which in a common Internet notation looks as "127.0.0.1:7171". After starting off our TCP server "sits" in a loop waiting for client requests. Once an incoming request has been detected, the TCP server creates a new communication channel frequently called a "working" socket which is used for data transfer between the server and the client. For a given example, data sent by the client is received by the server and output to screen. Upon completing the data transfer the server closes the working communication channel to the client and keeps on waiting for other requests.

The client application, from the programmer's point of view, is simpler than the TCP server – its main task is to pass data to the server. When data transfer has completed, the client application must close a working communication channel on its side.
Let's see the source code of both client and server. Below you can see a C source code of the TCP server application (Listing 5.1):

Listing 5.1

```
#include <stdio.h>
#include <stdlib.h>
#include <signal.h>
#include <unistd.h>
#include <sys/socket.h>
#include <resolv.h>
#include <string.h>

#define port 7171

int server;
pid_t PID;

void sig_handler(int signal_num)
{
  close(server);
  printf("Listening socket has closed.\n");
  kill(PID, SIGTERM);
}

int main(void)
{
```

```
  int client;
  struct sockaddr_in addr;
  char buf[4096];
  int bRead;
  struct sigaction sa;

  addr.sin_family = AF_INET;
  addr.sin_port = htons(port);
  addr.sin_addr.s_addr = inet_addr("127.0.0.1");

  server = socket(AF_INET, SOCK_STREAM, 0);
  bind(server, (struct sockaddr*)&addr, sizeof(addr));
  listen(server, 15);


  memset(&sa, 0, sizeof(sa));
  sa.sa_handler = &sig_handler;
  sigaction(SIGINT, &sa, NULL);
  PID = getpid();

  printf("TCP Demo Server waits for incoming requests on port 7171\
n");
  printf("Press Ctrl-C to terminate the program...\n");
  for(;;)
  {
    client = accept(server, 0 , 0);
    bRead = recv(client, buf, sizeof(buf), 0);
    if (bRead > 0)
    {
     buf[bRead] = '\0';
     printf("RECEIVED: %s\n", buf);
    }
    close(client);
  }
  close(server);
  return 0;
}
```

As we are dealing with sockets, we need to invoke some library functions referred to in a `sys/socket.h` header file. That is the reason we need to place the`include <sys/socket.h>` directive at the beginning of the source code. Often when writing a C network application it is convenient to operate with symbolic names instead of IP-addresses which are, in general, binary numbers. The C language comprises some library functions described in a `resolv.h` header file that implement conversions between native binary coded IP addresses and their symbolic representation, so we add the reference to this library in the source code.

In our program we defined `server` and `client` variables of an `int` type containing the descriptors of the listening and working sockets respectively. The `buf` variable will hold bytes received from the client in a current session.
Also we need to determine the structure named `addr` of a `sockaddr_in` type. This is a basic structure which must be filled with a basic parameters needed for configuring the listening socket. We need to configure `sin_addr`, `sin_port` and `sin_family` fields.
The program starts off by creating the listening socket:

```
server = socket(AF_INET, SOCK_STREAM, 0);
```

Initializing the server socket is provided by a `socket` library function taking three parameters. First on the left must be `AF_INET` (since we work with a TCP stack). With TCP the data is presented as a byte stream without losses, so the second parameter must be `SOCK_STREAM`. Finally, the last parameter that is on the right defines the type of protocol used. As the first two parameters completely define the type of protocol, we can assign zero to this parameter.
Note that for simplicity we don't process any errors that might appear when working with sockets. You must foresee similar situations and provide appropriate handling in those cases. Let's go back to our code.
After the socket has been successfully created we must bind its descriptor with the `addr` structure using the following statement:

```
bind(server, (struct sockaddr*)&addr, sizeof(addr));
```

The left parameter of the `bind()` function holds the descriptor of the listening socket (server, in the given case). The following parameter takes the address (pointer) of the `addr` structure that must have been already configured properly; the right parameter holds the size of the `addr` structure. After binding, the program puts the socket server into the listening state. The following statement does that:

```
listen(server, 15);
```

The `listen()` function sets the `server` socket to listen to incoming requests. The TCP server while terminated must close the descriptor of the listening socket. The right parameter, that is 15, determines the maximum number of working connections processing at any time.
By now we have finished the configuration of our listening socket. The next step is to arrange an endless loop in which our program will process the incoming requests from clients. Here, we use the `for()` loop. The loop begins with the

```
client = accept(server, 0 , 0);
```

statement that pauses (blocks) the program until the request from the client is delivered. In this case the `client` variable is assigned the descriptor of a new connection just established. Following this all operations will be processed using the client socket. After data transfer has completed the working connection must be deleted by using a `close()` function. All active connections will be closed anyway by the system after the program has ended, but I recommend you to close all descriptors when you no longer need them in your programs.
Assuming that the `client` socket has been created successfully, we read data prepared by the client using the `recv()` library function:

```
bRead = recv(client, buf, sizeof(buf), 0);
```

The `bRead` variable will contain a number of bytes received after `recv()` returns. If this value is not positive, you must handle it in your code. If `bRead` is greater than zero, this means that some data has been received, so we can process it. In our case we simply output data onto a screen by calling the `printf()` library function. Before outputting we need to place a terminated 0 after the last data byte in the `buf` variable thus creating a null-terminated string.
After displaying the string we have to close the working connection (the `client` socket):

```
close(client);
```

The above statement completes an iteration; the program goes back to the first statement of the `for()` loop and keeps on waiting for a new request. Our TCP server application will be ended after pressing Ctrl–C. As we need to correctly shutdown our server, we have configured an event handler responding to the Ctrl–C occurrence. If a user has pressed Ctrl–C, the `SIGINT` signal arises. To configure the SIGINT signal handler we place the following statements:

```
memset(&sa, 0, sizeof(sa));
sa.sa_handler = &sig_handler;
sigaction(SIGINT, &sa, NULL);
```

Upon pressing Ctrl-C the `SIGINT` signal is delivered to the process and the `sig_handler()` function is called. The first statement of `sig_handler()` closes the listening socket (the `server` variable), then sends a `kill` signal to the process by calling a `kill()` function:

```
kill(PID, SIGTERM);
```

The `kill()` call destroys the process with identifier PID indicated by the left parameter of this function. The `PID` variable must contain the process identifier after executing the following statement in the main program:

```
PID = getpid();
```

That's all about the TCP Server. Note that for brevity and clarity I don't process possible errors that the network library functions (socket, bind, accept, etc.) may return when completed. But if you want to write a reliable network application you must consider at least the most common situations that may cause errors (breaking a communication channel, a wrong IP-address or port, exceeding the limit of listening connections, etc.). I encourage you to explore the great mound of useful documentation available on this topic available online.

Now let's back to our network system. We will see how to implement a TCP client application operating with the server, described above. Imagine that the TCP client has to send some text string to the TCP server. Inmost aspects the TCP client functions like the server. The simplest client application should execute the following sequence:

1. create a socket through which the connection to the server will be established;
2. connect to the server via its IP-address / port. After successful establishing the connection the client may either send some data to the server or receive data from it;
3. close an active connection to the server thus terminating the communication.

In the client application we don't need to bind socket to the `sockaddr_in` structure as it takes place at the server, instead we must use the `connect()` function to establish a connection to the server. The C source code of a simple TCP client passing data to the server is shown in Listing 5.2:

Listing 5.2

```
#include <stdio.h>
#include <stdlib.h>
#include <unistd.h>
#include <sys/socket.h>
#include <resolv.h>
#include <string.h>
```

```
#define port 7171

int main(void)
{
 int client;
 struct sockaddr_in addr;
 char *str = "Test string for a TCP server";

 addr.sin_family = AF_INET;
 addr.sin_port = htons(port);
 addr.sin_addr.s_addr = inet_addr("127.0.0.1");

 client = socket(AF_INET, SOCK_STREAM, 0);
 connect(client, (struct sockaddr*)&addr, sizeof(addr));
 send(client, str, strlen(str), 0);
 close(client);
 return 0;
 }
```

As in the case of the server, we need to create the `sockaddr_in` structure (named as `addr`) and fill it with proper values. Our server and client will be running on the same PC, so we put "127.0.0.1" in the `sin_addr.s_addr` field; the `sin_port` field must contain the 7171 that coincides with the number of the listening port of the server. The `sin_family` is assigned the `AF_INET` constant that identifies the TCP applications. Then the program creates a socket whose descriptor is held in the `client` variable. The next step is to establish a connection with the server using a `connect()` library function:

```
connect(client, (struct sockaddr*)&addr, sizeof(addr));
```

As you can see the `connect()` function takes the same parameters as `bind()` used in the TCP server. Supposing the connection has been successfully established we send the `str` text string to the server:

```
send(client, str, strlen(str), 0);
```

Finally, the client socket is closed by a `close(client)` function.
Suppose we put the C source code of the server in the DemoTCPServer.c and the client source code in DemoTCPClient.c. To compile both files we can enter the following command strings:

```
$ gcc -o DemoTCPServer DemoTCPServer.c
$ gcc -o DemoTCPClient DemoTCPClient.c
```

After successful compilation we may launch both executables, the DemoTCPServer application first. After the TCP server has started we can execute the client. Below you see two terminal windows with the server (Fig.5.5) and the client (Fig.5.6) running.

```
File  Edit  View  Terminal  Help

yury@yury-desktop:~/Networking$ ./DemoTCPServer
TCP Demo Server waits for incoming requests on port 7171
Press Ctrl-C to terminate the program...
RECEIVED: Test string for a TCP server
RECEIVED: Test string for a TCP server
RECEIVED: Test string for a TCP server
^CListening socket has closed.
Terminated
yury@yury-desktop:~/Networking$ 
```

Fig.5.5

As you can see, the TCP server has terminated after receiving three text strings from the client.

```
File  Edit  View  Terminal  Help
yury@yury-desktop:~/Networking$ ./DemoTCPClient
yury@yury-desktop:~/Networking$ ./DemoTCPClient
yury@yury-desktop:~/Networking$ ./DemoTCPClient
yury@yury-desktop:~/Networking$
yury@yury-desktop:~/Networking$ netstat -l | grep localhost
tcp        0      0 localhost:7171          *:*                     LISTEN
tcp        0      0 localhost:ipp           *:*                     LISTEN
tcp6       0      0 localhost:ipp           [::]:*                  LISTEN
yury@yury-desktop:~/Networking$
yury@yury-desktop:~/Networking$
yury@yury-desktop:~/Networking$ netstat | grep localhost
tcp        0      0 localhost:36213         localhost:7171          TIME_WAIT
tcp        0      0 localhost:36214         localhost:7171          TIME_WAIT
tcp        0      0 localhost:36212         localhost:7171          TIME_WAIT
yury@yury-desktop:~/Networking$
```

Fig.5.6

The Linux operating system allows us to configure and view the network interaction by using some powerful commands. One of them, `netstat`, allows us to observe the current state of the network communications including IP-addresses, ports and connection activities. As Fig.5.6 implies, the state of our TCP server may be revealed by entering the following two commands:

```
$ netstat -l | grep localhost
$ netstat | grep localhost
```

As the IP loopback address 127.0.0.1 in Linux is associated with the name "`localhost`", we can use this name in network commands. The first command taken with a `-l` parameter displays the state of the listening socket waiting for requests from clients.
The second `netstat` command typed in without parameters exhibits the active connections to the server. You can see that there were three active connections opened on ports 36213, 36214 and 36212. Each of these ports was allocated dynamically by the system after a new working connection had been accepted. The `netstat` command has a lot of other parameters which provide you more various and useful information concerning network activity – you may view Linux documentation for further details.
The above client–server example illustrates a one–way data transfer from the client to the server. The client may also receive data from the server or implement data transfer in both directions. As we want to apply networking to control hardware ports, the client that receives data from the server comes in handy when we need, say, to check the state of some hardware register by periodically polling the server which, in turn, reads a given register.
To realize such a client-server configuration we have to make some changes to both client and server source codes. Below you can see the C source code of the TCP server which sends some data back to the TCP client (Listing 5.3):

Listing 5.3

```
#include <stdio.h>
#include <stdlib.h>
#include <signal.h>
#include <unistd.h>
#include <sys/socket.h>
#include <resolv.h>
#include <string.h>


#define port 7171

int server;
pid_t PID;

void sig_handler(int signal_num)
{
 close(server);
```

```
printf("Listening socket has closed.\n");
kill(PID, SIGTERM);
}

int main(void)
{
 int client;
 struct sockaddr_in addr;
 char *str = "String sent by the TCP server.";
 struct sigaction sa;

 addr.sin_family = AF_INET;
 addr.sin_port = htons(port);
 addr.sin_addr.s_addr = inet_addr("127.0.0.1");

 server = socket(AF_INET, SOCK_STREAM, 0);
 bind(server, (struct sockaddr*)&addr, sizeof(addr));
 listen(server, 15);


 memset(&sa, 0, sizeof(sa));
 sa.sa_handler = &sig_handler;
 sigaction(SIGINT, &sa, NULL);
 PID = getpid();


 printf("TCP Demo Server waits for incoming requests on port 7171\
n");
 printf("Press Ctrl-C to terminate the program...\n");
 for(;;)
 {
  client = accept(server, 0 , 0);
  send(client, str, strlen(str), 0);
  close(client);
 }
 close(server);
 return 0;
}
```

As you can see, the C source code of this server is almost the same as that shown in Listing 5.1. The only difference is that the server sends a text string `str` back to the client instead of receiving data from the client. To send data to the client we replaced the `recv()` call (see Listing 5.1) with `send()` in a `for()` loop:

```
for(;;)
{
 client = accept(server, 0 , 0);
 send(client, str, strlen(str), 0);
 close(client);
}
```

The C source code of the TCP client is based on that from Listing 5.2 except with some changes. The modified version of the TCP client source code is shown in Listing 5.4:

Listing 5.4

```
#include <stdio.h>
#include <stdlib.h>
#include <unistd.h>
#include <sys/socket.h>
#include <resolv.h>
#include <string.h>


#define port 7171

int main(void)
{
 int client;
 struct sockaddr_in addr;
 char buf[4096];
 int bRead;

 addr.sin_family = AF_INET;
 addr.sin_port = htons(port);
 addr.sin_addr.s_addr = inet_addr("127.0.0.1");
```

```
 client = socket(AF_INET, SOCK_STREAM, 0);
 connect(client, (struct sockaddr*)&addr, sizeof(addr));
 bRead = recv(client, buf, sizeof(buf), 0);
 if (bRead > 0)
  {
   buf[bRead] = '\0';
   printf("CLIENT RECEIVED: %s\n", buf);
  }
 close(client);
 return 0;
}
```

As our client must receive the byte stream, the following section was introduced:

```
bRead = recv(client, buf, sizeof(buf), 0);
if (bRead > 0)
{
 buf[bRead] = '\0';
 printf("CLIENT RECEIVED: %s\n", buf);
}
```

After a working connection has been established the TCP client receives data from the server using the `recv()` library function. If `bRead` is greater than 0 (some data is received) the content of `buf` is displayed onto the screen.

Internet browser as a network client

Thus far we have considered distributed applications with the server and the client written in C. Many times we can simplify the development by taking some popular Web–browser which plays the role of a TCP client. Additionally, using a Web-browser saves a lot of time since we don't need to write any client source code any more. We can take any popular web-browser and use it for receiving data from some TCP or HTTP server. As the HTTP protocol is based on the TCP protocol suite we will use terms "TCP server" and "HTTP server" as synonyms. In fact, a Web-browser is a TCP client that sends to the TCP server a request that is formed using HTTP protocol specifications.

We will not consider HTTP protocol in depth; you can find lots of documents, literature and tutorials on this topic on the Internet with just a little research. We will only discuss practical aspects of designing the systems "HTTP server – Web-browser". To see how such a distributed system will

function we will take a simple HTTP server written in C and some popular browser (I use Mozilla FireFox, although you may utilize another popular Web-browser if needed).
Let's go to the practice and begin with the C source code of a simple HTTP server (Listing 5.5):

Listing 5.5

```
#include <stdio.h>
#include <stdlib.h>
#include <signal.h>
#include <unistd.h>
#include <sys/socket.h>
#include <resolv.h>
#include <string.h>

#define port 8181

int server;
pid_t PID;

void sig_handler(int signal_num)
{
  close(server);
  printf("Listening socket has closed.\n");
  kill(PID, SIGTERM);
}

int main(void)
{
 int client;
 struct sockaddr_in addr;
 char buf[4096];
 char *answ = "<html><body><h2> HTTP Server answered OK. </h2></
body></html>";
 int bRead;

 struct sigaction sa;

 addr.sin_family = AF_INET;
 addr.sin_port = htons(port);
```

```
 addr.sin_addr.s_addr = inet_addr("127.0.0.1");

 server = socket(AF_INET, SOCK_STREAM, 0);
 bind(server, (struct sockaddr*)&addr, sizeof(addr));
 listen(server, 15);

 memset(&sa, 0, sizeof(sa));
 sa.sa_handler = &sig_handler;
 sigaction(SIGINT, &sa, NULL);
 PID = getpid();
 printf("HTTP Demo Server waits for incoming requests on port 8181\
n");
 printf("Press Ctrl-C to terminate the program...\n");
 for(;;)
 {
   client = accept(server, 0 , 0);
   bRead = recv(client, buf, sizeof(buf), 0);
   if (bRead > 0)
   {
    buf[bRead] = '\0';
    printf("RECEIVED: %s\n", buf);
    send(client, answ, strlen(answ), 0);
   }
   close(client);
 }
 close(server);
 return 0;
}
```

Let's recall that the Web-browser basically handles a HTML page which will be sent to it by a HTTP server in response to the browser's HTTP request. This means that our HTTP server must send either a HTML page saved in .htm or .html file or the HTML page created"on the fly". For clarity we created our HTML page inside the source code by introducing the `answ` string. This string is sent to the web-browser after a new HTTP request has arrived.

Our HTTP server sits in an infinite `for()` loop listening to port 8181. When a new request has been detected the server code opens a new working connection (the `client` socket) through which the HTML page held in the `answ` variable is sent to the browser.

Below you can see a browser's window after sending the request to the HTTP server (Fig.5.7):

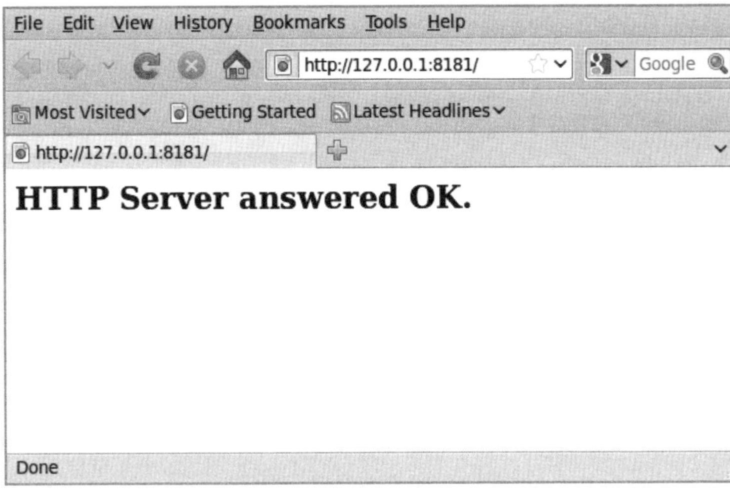

Fig.5.7

Our browser has generated a standard GET request as is shown below:

```
RECEIVED: GET /favicon.ico HTTP/1.1
Host: 127.0.0.1:8181
User-Agent: Mozilla/5.0 (X11; U; Linux i686; en-US; rv:1.9.1.7)
Gecko/20100106 Ubuntu/9.10 (karmic) Firefox/3.5.7
Accept: text/html,application/xhtml+xml,application/xml;q=0.9,*/
*;q=0.8
Accept-Language: en-us,en;q=0.5
Accept-Encoding: gzip,deflate
Accept-Charset: ISO-8859-1,utf-8;q=0.7,*;q=0.7
Keep-Alive: 300
Connection: keep alive
```

This request is seen in the terminal window where the HTTP server has been launched (Fig.5.8):

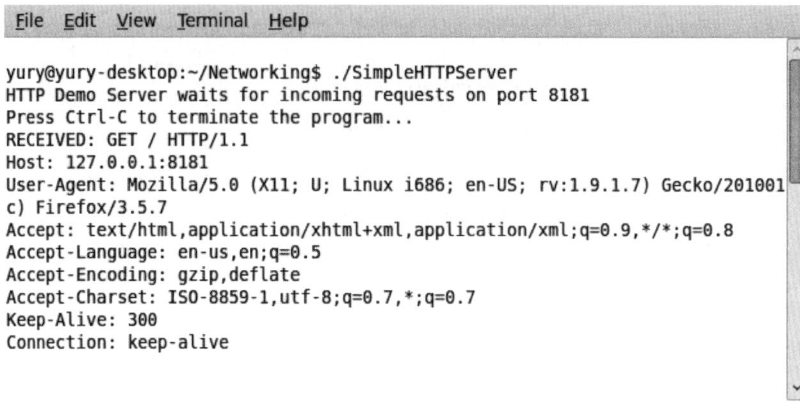

Fig.5.8

As you can see from the above screenshot, the browser issued a standard GET request which our HTTP server processed as shown in Fig.5.8.

Hardware access from network applications

By now we know how to build a simple "client-server" distributed system based on TCP/IP network protocol suite. Using this programming architecture we can easily design embedded systems capable of implementing a remote control or a data acquisition. Several demo projects described below show practical aspects in realizing such distributed control systems.

Suppose we want to write a byte to the hardware port on the local PC from a remote host. Let's select the data port of the standard parallel interface with an address 0x378 (you should select the appropriate address assigned to the data port of your parallel interface). We will test such a system on the same machine, so the IP-address will be 127.0.0.1 for both server and client applications.
Our distributed system will operate in the following way: the TCP client sends a data byte to the server that sits on port 7171 waiting for requests. The server, after accepting the request receives a byte from the client and writes it to the data port. This, in brief, is how our system works.
Let's begin with the C source code for the TCP server (Listing 5.6):

Listing 5.6

```
#include <stdio.h>
#include <sys/io.h>
#include <stdlib.h>
#include <signal.h>
#include <unistd.h>
#include <sys/socket.h>
#include <resolv.h>
#include <string.h>

#define port 7171

int server;
pid_t PID;
void sig_handler(int signal_num)
{
  close(server);
  printf("Listening socket has closed.\n");
  kill(PID, SIGTERM);
}

int main(void)
{
 int client;
 struct sockaddr_in addr;
 char bToWrite;
 int bRead;

 struct sigaction sa;
 ioperm(0x378, 2, 1);
 addr.sin_family = AF_INET;
 addr.sin_port = htons(port);
 addr.sin_addr.s_addr = inet_addr("127.0.0.1");
 server = socket(AF_INET, SOCK_STREAM, 0);
 bind(server, (struct sockaddr*)&addr, sizeof(addr));
 listen(server, 15);

 memset(&sa, 0, sizeof(sa));
 sa.sa_handler = &sig_handler;
 sigaction(SIGINT, &sa, NULL);
```

```
 PID = getpid();

 printf("TCP Server waits for the byte to write to address 0x378\
n");
 printf("Press Ctrl-C to terminate the program...\n");
 for(;;)
 {
   client = accept(server, 0 , 0);
   bRead = recv(client, &bToWrite, sizeof(char), 0);
   if (bRead > 0)
   {
    printf("BYTE TO WRITE: 0x%X\n", bToWrite);
    outb(bToWrite, 0x378);
   }
   close(client);
 }
 close(server);
 return 0;
}
```

Let's discuss this source code. The major portion of it is already realized for you from the previous examples, so we highlight only the section where hardware is involved.
Recall that dealing with hardware requires use of `ioperm()`, `iopl()`, `inb()` and `outb()` functions declared in the `sys/io.h` header file, so we must declare this file in the `include` directive at the beginning of the listing. The byte for writing to the data port is held in the `bWrite` variable of a `char` type. Again, before any operations with hardware ports the process must gain appropriate permissions, so we introduced the `ioperm(0x378, 2, 1)` function.
Receiving byte from the client and writing it to the data port is accomplished thus:

```
 . . .
 bRead = recv(client, &bToWrite, sizeof(char), 0);
 if (bRead > 0)
 {
  printf("BYTE TO WRITE: 0x%X\n", bToWrite);
  outb(bToWrite, 0x378);
 }
 . . .
```

The `recv()` function tries to receive a data byte and place it in the `bToWrite` variable. If successful (`bRead` is greater than 0) the value in `bToWrite` is displayed in the terminal window and written to the data port using the `outb (bToWrite, 0x378)` function.

The TCP client as you may guess is simple and straightforward. The program code takes a byte provided as a parameter upon launching the client application and sends it to the TCP server. The source code of the client is shown below (Listing 5.7):

Listing 5.7

```
#include <stdio.h>
#include <stdlib.h>
#include <unistd.h>
#include <sys/socket.h>
#include <resolv.h>
#include <string.h>

#define port 7171

int main(int argc, char *argv[])
{
  int client;
  struct sockaddr_in addr;
  int num;
  char bToWrite;

  if (argc != 2)
  {
   printf("Usage: %s byte\n", argv[0]);
   return 0;
  }

 num = atoi(argv[1]);
 bToWrite = (char)num;
 addr.sin_family = AF_INET;
 addr.sin_port = htons(port);
 addr.sin_addr.s_addr = inet_addr("127.0.0.1");

 client = socket(AF_INET, SOCK_STREAM, 0);
 connect(client, (struct sockaddr*)&addr, sizeof(addr));
 send(client, &bToWrite, 1, 0);
 close(client);
 return 0;
}
```

The byte the client sends to the server is held in the `bToWrite` variable of a `char` type. After creating a socket with the `client` descriptor and establishing the connection to the server the program sends the byte using the `send()` call.
You can see the terminal window where the server application is executing in Fig.5.9:

```
File  Edit  View  Terminal  Help
yury@yury-desktop:~/Networking$ gcc -o TCPServer_WriteByte TCPServer_WriteByte.c
yury@yury-desktop:~/Networking$ clear
yury@yury-desktop:~/Networking$ gcc -o TCPServer_WriteByte TCPServer_WriteByte.c
yury@yury-desktop:~/Networking$ sudo ./TCPServer_WriteByte
TCP Server waits for the byte to write to the address 0x378
Press Ctrl-C to terminate the program...
BYTE TO WRITE: 0x0
BYTE TO WRITE: 0x1
BYTE TO WRITE: 0x2
BYTE TO WRITE: 0x3
^CListening socket has closed.
Terminated
yury@yury-desktop:~/Networking$
```

Fig.5.9

As you can see, the TCP server received four bytes with values beginning from 0x0 in succession and wrote them to port 0x378. To test the data port you may measure the voltage swing on pins 2–3 (bytes 0 – 1 of the port 0x378) of the DB-25 connector of the parallel interface.
The terminal window of the client application is shown in Fig.5.10:

```
File  Edit  View  Terminal  Help

yury@yury-desktop:~/Networking$ ./TCPClient_WriteByte 0
yury@yury-desktop:~/Networking$ ./TCPClient_WriteByte 1
yury@yury-desktop:~/Networking$ ./TCPClient_WriteByte 2
yury@yury-desktop:~/Networking$ ./TCPClient_WriteByte 3
yury@yury-desktop:~/Networking$
```

Fig.5.10

Reading hardware ports over a network may be implemented after we slightly modify both TCP client and server applications discussed above. In this configuration the client application issues the request to the TCP server which, in turn, returns the byte read from some hardware port to the client. Our next project demonstrates reading a byte from the data port of the parallel interface and sending it to the TCP client. Let's begin at the TCP server with the source code given below (Listing 5.8):

Listing 5.8

```
#include <stdio.h>
#include <sys/io.h>
#include <stdlib.h>
#include <signal.h>
#include <unistd.h>
#include <sys/socket.h>
#include <resolv.h>
#include <string.h>

#define port 7171

int server;
pid_t PID;

void sig_handler(int signal_num)
{
  close(server);
  printf("A listening socket has closed.\n");
  kill(PID, SIGTERM);
}

int main(void)
{
 int client;
 struct sockaddr_in addr;

 char bRead;
 char test = 0;
 struct sigaction sa;

 ioperm(0x378, 2, 1);

 addr.sin_family = AF_INET;
 addr.sin_port = htons(port);
 addr.sin_addr.s_addr = inet_addr("127.0.0.1");

 server = socket(AF_INET, SOCK_STREAM, 0);
 bind(server, (struct sockaddr*)&addr, sizeof(addr));
 listen(server, 15);
```

```
 memset(&sa, 0, sizeof(sa));
 sa.sa_handler = &sig_handler;
 sigaction(SIGINT, &sa, NULL);
 PID = getpid();

 printf("TCP Server waits for the byte to read from 0x378\n");
 printf("Press Ctrl-C to terminate the program...\n");
 for(;;)
 {
   client = accept(server, 0 , 0);
   outb(test, 0x378);
   bRead = inb(0x378);
   test++;
   send(client, &bRead, sizeof(char), 0);
   close(client);
 }

 close(server);
 return 0;
}
```

As usual, the TCP server waits for incoming requests on port 7171; after the client has established the connection the server reads a byte from the data port and sends it back to the client. To provide a quick view of results I applied the following trick:

As the data port may be written to and read from, we can write some value to the data port and immediately read this value back. We use a `test` variable for that purpose. The `test` variable is written to the data port; we must get this value back a little later when reading a byte from the data port in the `bRead` variable. Then the `test` variable is incremented;preparing for next use. This sequence allows us to see how the result changes after some requests from the TCP client to the server. The byte held in `bRead` then is sent back to the client.

This sequence is realized in the following section of the source code:

```
outb(test, 0x378);
bRead = inb(0x378);
test++;
send(client, &bRead, sizeof(char), 0);
```

That's all about the server. The TCP client simply connects to the server through port 7171, gains the byte read and displays it onto screen. The source code of the client is shown below (Listing 5.9):

Listing 5.9

```
#include <stdio.h>
#include <stdlib.h>
#include <unistd.h>
#include <sys/socket.h>
#include <resolv.h>
#include <string.h>

#define port 7171

int main(void)
{
 int client;
 struct sockaddr_in addr;
 char bRead;

 addr.sin_family = AF_INET;
 addr.sin_port = htons(port);
 addr.sin_addr.s_addr = inet_addr("127.0.0.1");

 client = socket(AF_INET, SOCK_STREAM, 0);

 connect(client, (struct sockaddr*)&addr, sizeof(addr));
 recv(client, &bRead, 1, 0);
 close(client);
 printf("BYTE READ: 0x%X\n", bRead);
 return 0;
}
```

The source code of the client is extremely simple. As usual, the program creates the socket `client` for accessing to the TCP server. After connection to the server has been established the `client` descriptor handles the received data byte using the `recv()` function and then displays this value onto the screen via the `printf()` library function.
Recall, that in real life when a TCP client and server are running on the different PCs you need to correctly set IP-addresses in their source codes.

We can simplify the above distributed system if we replace our client application with a Web–browser. We have already discussed the example where both the HTTP server and the Web–browser communicate with each other. Now let's make a useful practical project where the HTTP server will read the data port (0x378) and send data to the Web-browser. As usual, we can test both server and client on the same machine using a loopback interface with IP-address 127.0.0.1. In this project we will set port 8181 for receiving incoming requests from clients.

Below you can see a source code of the HTTP server (Listing 5.10):

Listing 5.10

```
#include <stdio.h>
#include <sys/io.h>
#include <stdlib.h>
#include <signal.h>
#include <unistd.h>
#include <sys/socket.h>
#include <resolv.h>
#include <string.h>

#define port 8181

int server;
pid_t PID;

void sig_handler(int signal_num)
{
  close(server);
  printf("Listening socket has closed.\n");
  kill(PID, SIGTERM);
}

int main(void)
{
 int client;
 struct sockaddr_in addr;
 char *str1 = "<html><body><h2> The data port (0x378) state: 0x";
 char bVal;
 char *str2 = " </h2></body></html>";
 char buf[256];
 int bRead;

 memset(buf, '\0', sizeof(buf));
```

```
char test = 0;
struct sigaction sa;

ioperm(0x378, 2, 1);

addr.sin_family = AF_INET;
addr.sin_port = htons(port);
addr.sin_addr.s_addr = inet_addr("127.0.0.1");

server = socket(AF_INET, SOCK_STREAM, 0);
bind(server, (struct sockaddr*)&addr, sizeof(addr));
listen(server, 15);
memset(&sa, 0, sizeof(sa));
sa.sa_handler = &sig_handler;
sigaction(SIGINT, &sa, NULL);
PID = getpid();

printf("HTTP Server is ready to read byte from address 0x378\n");
printf("Press Ctrl-C to terminate the program...\n");

for(;;)
{
  client = accept(server, 0 , 0);
  if (client > 0)
   {
    outb(test, 0x378);
    test++;
    bVal = inb(0x378);

  // Creating message

    memset(buf, '\0', sizeof(buf));
    bRead = sprintf(buf, "%s", str1);
    bRead += sprintf(buf + bRead, "%X", bVal);
    bRead += sprintf(buf + bRead, "%s", str2);

  // Sending message to the client

   write(client, buf, bRead);
   }
```

```
    close(client);
  }
 close(server);
 return 0;
}
```

The source code of the TCP server is already known to us except some new sections which we will discuss. The following section realizes the algorithm already discussed (see Listing 5.8):

```
outb(test, 0x378);
test++;
bVal = inb(0x378);
```

After reading a byte from the data port into `bVal` the program code creates a text string to send to the browser using the `sprint()` C library function. This function fills the `buf` array by concatenating formatted values as shown in the following section of code:

```
bRead = sprintf(buf, "%s", str1);
bRead += sprintf(buf + bRead, "%X", bVal);
bRead += sprintf(buf + bRead, "%s", str2);
```

The resulting string consisting of `str1`, `bVal` and `str2` is placed in the `buf` array, so `buf` will lastly contain a simple HTML page. The `bRead` variable will hold the length of the string stored in `buf`. The following statement sends this buffer to the browser:

```
write(client, buf, bRead);
```

Note that we use a `write()` system call instead of the `send()` network library function – we can do this for the `client` socket. It is an object belonging to the file system and we can apply a low–level `write()` system call to this object. The same is true for the `read()` system call that can be used instead of a `recv()` network library function. Finally we end operations on the working connection by closing the `client` socket:

```
close(client);
```

Let's save the above source code in the HTTPEmbedServer.c file and start the compilation. If there are no errors in the source code we will gain an executable named HTTPEmbedServer. As our program tries to access hardware, we must launch our server application using `sudo` command. Below you can see the terminal window where the server is running (Fig.5.11):

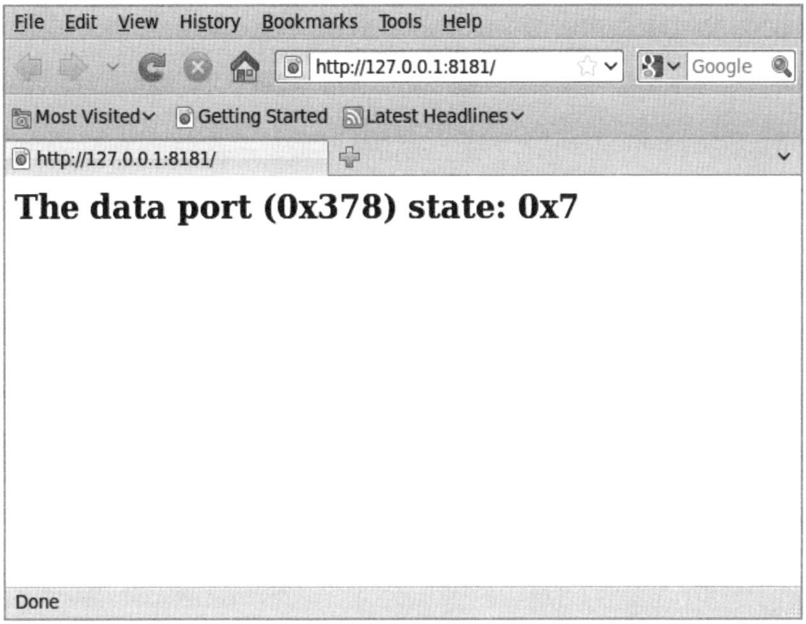

Fig.5.11

As usual, terminating the server is implemented after pressing Ctrl–C. The following graphics displays a Web-browser's window after a HTTP request to the server has been performed (Fig.5.12):

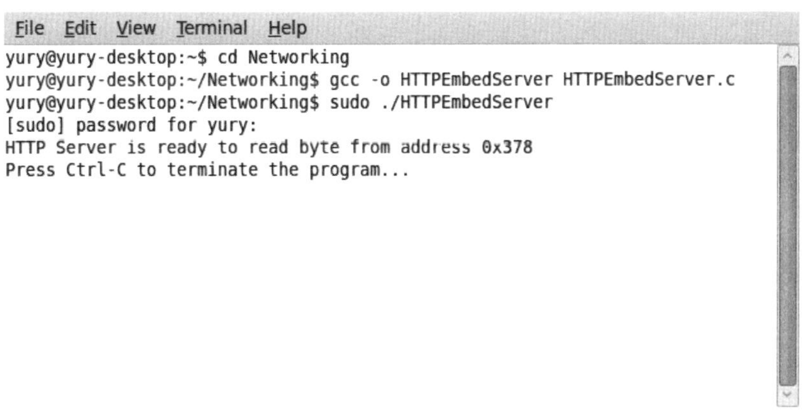

Fig.5.12

If we issue the same HTTP request once more the state of the data port reflected in the Web-browser's window will be 0x8. The next request results in 0x9 etc. When accessing the HTTP server you can also enter the "localhost" instead of "127.0.0.1" in the address field of the Web-browser, so the URL might look like `http://localhost:8181`.

Recall that besides TCP/IP networking the Linux operating system has another useful feature for designing distributed applications known as "local socket". The next section describes a practical example of using local sockets.

Client–Server applications with local sockets

As we already know, local sockets may come in handy when we need to develop a distributed system consisting of some applications running on the same machine. For a better understanding of local sockets let's discuss a simple client–server system. Below you can see the C source code of a local server which waits for incoming requests from a local client, then receives data and displays it onto screen (Listing 5.11):

Listing 5.11

```
#include <unistd.h>
#include <sys/un.h>
#include <sys/socket.h>
#include <stdio.h>

int main(void)
{
 int localServer, localClient;
 struct sockaddr_un addr;
 char buf[4096];
 int bRead;
 unlink("/home/yury/LOCAL_SOCKETS/mysocket");

 localServer = socket(AF_LOCAL, SOCK_STREAM, 0);
 bzero(&addr, sizeof(addr));
 addr.sun_family = AF_LOCAL;
 strcpy(addr.sun_path, "/home/yury/LOCAL_SOCKETS/mysocket");
 bind(localServer, (struct sockaddr*)&addr, sizeof(addr));
 listen(localServer, 5);
 printf("Local Server waits for incoming requests...\n");
 while(1)
```

```
 {
   localClient = accept(localServer, 0, 0);
   if (localClient > 0)
   {
    bRead = recv(localClient, buf, sizeof(buf), 0);
    if (bRead > 0)
     {
      buf[bRead] = '\0';
      printf("RECEIVED: %s\n", buf);
     }
   }
  close(localClient);
  }
 close(localServer);
 return 0;
}
```

As we are dealing with local sockets, we need to place the `include` directive to refer to the `sys/un.h` header file at the beginning of the source code. We also defined two variables `localServer` and `localClient` – they will hold the descriptor of the listening socket and the descriptor of the current working connection respectively. The `buf` variable will contain data received through an active connection and the `bRead` variable will hold the length of data received. Also we define the `addr` structure of `struct sockaddr_un` type which is needed for configuring the listening socket.

Our program creates the local socket named "`mysocket`" – the server will take incoming requests there. Before creating "`mysocket`" we must delete the local socket with the same path name if it already exists. We do that using the following statement:

```
unlink("/home/yury/LOCAL_SOCKETS/mysocket");
```

The `unlink()` call deletes the object of the file system indicated by a single parameter. If a given object doesn't exist, `unlink()` does nothing and returns an error. Anyway, after this function has been executed we can be certain that the local socket "`mysocket`" doesn't exist anymore.

The statement that follows creates the `localServer` listening socket:

```
localServer = socket(AF_LOCAL, SOCK_STREAM, 0);
```

As in the case of using TCP sockets, the `socket()` function is applied here, but one of its parameters should be the AF_LOCAL constant instead of AF_INET. Our `localServer` socket must be bound to the `addr` structure, so we need to fill `sun_family` and `sun_path` fields with appropriate values:

```
addr.sun_family = AF_LOCAL;
strcpy(addr.sun_path, "/home/yury/LOCAL_SOCKETS/mysocket");
```

The `sun_family` field is assigned `AF_LOCAL` and `sun_path` contains the full path name to the local socket. The next statement makes the binding:

```
bind(localServer, (struct sockaddr*)&addr, sizeof(addr));
```

Now we can put `localServer` into the listening state until the application terminates. Our server application enters an endless `while()` loop waiting for incoming requests. The first statement in the `while()` loop is the `accept()` function that allows creation of a new working connection (socket) with the `localClient` descriptor:

```
localClient = accept(localServer, 0, 0);
```

If the new connection has been successfully established (the `localClient > 0` condition) the program receives a byte stream using the `recv()` function and places data in the memory buffer `buf`. Finally, the content of `buf` is displayed onto the screen. Since the program doesn't need this working connection anymore, the `close(localClient)` statement closes the `localClient` socket and goes to the start point of the `while()` loop.

This in brief is how this server functions. The terminal window where the local server application is running should look something like that shown in Fig.5.13:

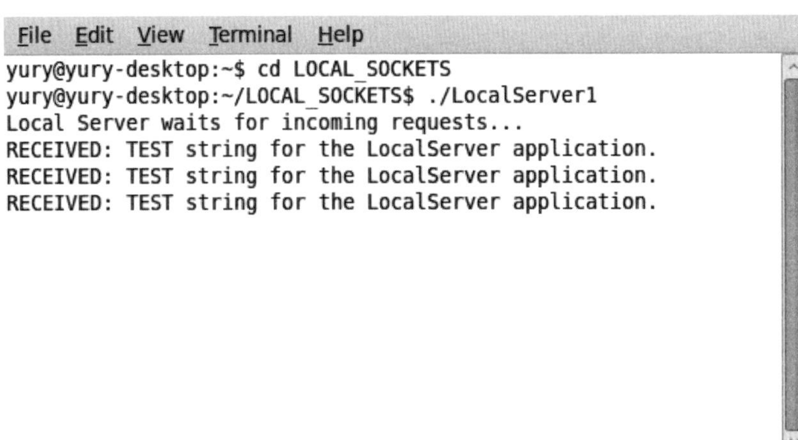

Fig.5.13

The local client application with source codeshown in Listing 5.12 simply sends the text string to the server:

Listing 5.12

```
#include <unistd.h>
#include <sys/un.h>
#include <sys/socket.h>
#include <stdio.h>
#include <string.h>

int main(void)
{
 int localClient;
 struct sockaddr_un addr;
 char *str = "TEST string for the LocalServer application.";

 localClient = socket(AF_LOCAL, SOCK_STREAM, 0);
 bzero(&addr, sizeof(addr));
 addr.sun_family = AF_LOCAL;
 strcpy(addr.sun_path, "/home/yury/LOCAL_SOCKETS/mysocket");
 connect(localClient, (struct sockaddr*)&addr, sizeof(addr));
 send(localClient, str, strlen(str), 0);
 close(localClient);
 return 0;
}
```

As in the case of the server, the client source code must contain the `include` directive referring to by the `sys/un.h` header file. We need only a single local socket here (`localClient`) for establishing the connection and sending data through the communication channel. The program initializes the `addr` structure with proper values which must coincide with those for the server application.

The terminal window where the client application is running is shown in Fig.5.14:

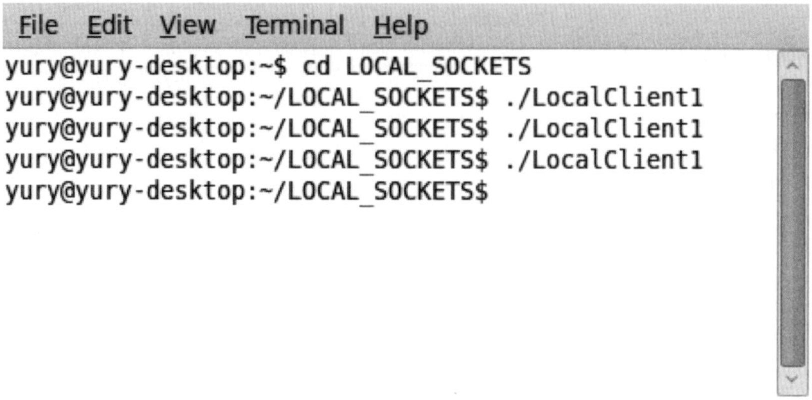

Fig.5.14

When dealing with local sockets we can check the state of the connection using the `netstat` command discussed earlier in this chapter. Fig.5.15 illustrates a result that `netstat` returns for our "`mysocket`" local socket:

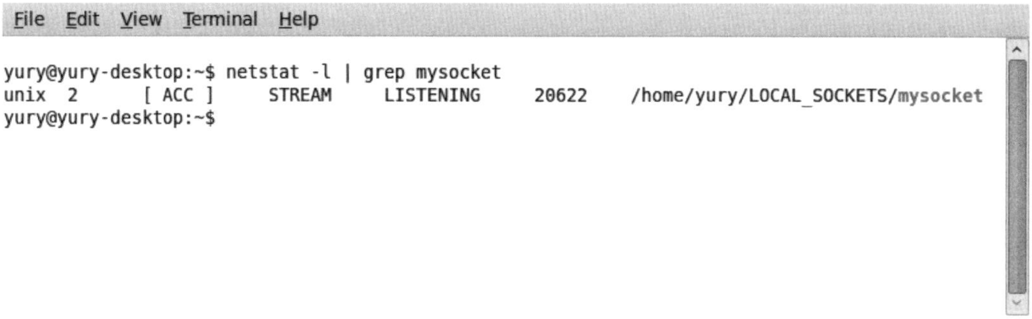

Fig.5.15

By now we already know how to operate with local sockets, so let's see a simple distributed system where we touch hardware.

Accessing hardware using local sockets

Our system consists of two standalone applications (client and server) running on the same machine. The local server waits for incoming requests from the client which wants to know the state of data port (address 0x378) of the parallel interface. When this request has been accepted the server opens a new working connection to the client, reads the data byte from the data port and sends information gained back to the client. Both client and server use the "`mysocket`" local socket for interaction. The C source code of the local server is shown in Listing 5.13:

Listing 5.13

```
#include <unistd.h>
#include <sys/un.h>
#include <sys/socket.h>
#include <sys/io.h>
#include <stdio.h>
#include <strings.h>

int main(void)
{
 int localServer, localClient;
 struct sockaddr_un addr;

 int bRead;
 char bVal;
 char test = 0x1;
 unlink("/home/yury/LOCAL_SOCKETS/mysocket");
 iopl(3);

 localServer = socket(AF_LOCAL, SOCK_STREAM, 0);
 bzero(&addr, sizeof(addr));
 addr.sun_family = AF_LOCAL;
 strcpy(addr.sun_path, "/home/yury/LOCAL_SOCKETS/mysocket");
 bind(localServer, (struct sockaddr*)&addr, sizeof(addr));
 listen(localServer, 5);
 printf("Checking state of a data port (0x378)...\n");
```

```
  while(1)
  {
    localClient = accept(localServer, 0, 0);
    if (localClient > 0)
     {
      outb(test, 0x378);
      test++;
      bVal = inb(0x378);
      send(localClient, &bVal, sizeof(char), 0);
     }
    close(localClient);
  }
  close(localServer);
  return 0;
}
```

As we are already familiar with the server application based on local sockets, let's consider what is new in this application. Our server needs to access hardware with `iopl()`, `inb()` and `outb()` privileged instructions, so we must declare these functions through `sys/io.h` header file using the `include` directive at the beginning of the source code. Then we grant an access to I/O ports by introducing the `iopl(3)` statement before touching hardware. To illustrate the work of the server the program code writes the value of the `test` variable to the data port, then reads the same value into the `bVal` variable and sends `bVal` to the local client. This approach was already discussed when analyzing Listing 5.8.
The local client is much simpler (Listing 5.14):

Listing 5.14

```
#include <unistd.h>
#include <sys/un.h>
#include <sys/socket.h>
#include <stdio.h>
#include <string.h>

int main(void)
{
 int localClient;
 struct sockaddr_un addr;
 char bVal;
 int bRead;
```

```
 localClient = socket(AF_LOCAL, SOCK_STREAM, 0);
 bzero(&addr, sizeof(addr));
 addr.sun_family = AF_LOCAL;
 strcpy(addr.sun_path, "/home/yury/LOCAL_SOCKETS/mysocket");
 connect(localClient, (struct sockaddr*)&addr, sizeof(addr));

 bRead = recv(localClient, &bVal, sizeof(char), 0);
 if (bRead > 0)
  printf("Byte read from 0x378: 0x%X\n", bVal);
 close(localClient);
 return 0;
}
```

The client application after establishing the connection to the server tries to read data into the `bVal` variable. If the operation was successful `bVal` is displayed onto the screen.

Designing distributed applications with MATLAB

This section is dedicated to making distributed applications using MATLAB. As an advanced scientific and engineering tool, MATLAB provides support for networking with TCP and UDP protocols. We will consider the basic principles of designing TCP/IP based distributed applications in the MATLAB environment. If you experiment with networking in MATLAB you may notice that designing network applications in that environment is simple and straightforward. This is because MATLAB operates with high-level library functions especially developed to simplify network programming. On the other hand, it is difficult to provide full functionality and high flexibility using the MATLAB network functions, as could be achieved when writing network applications, say, in C or Python. Nevertheless, many times it would be wise to use MATLAB network features for data transfer between applications and interaction applications with each other.
Being a very powerful engineering and scientific tool, MATLAB is often configured as a network client.

When we start building a network client application in MATLAB, we first need to create a TCP/IP object by using the `tcpip()` function. This function usually takes two input arguments, the name of the remote host and a remote port value. If port 80 is used the `tcpip()` function may take a single argument that is the name of the remote host. In our projects we will also configure some common parameters of a TCP/IP object created by `tcpip()`. If you want to know more about setting parameters for TCP/IP objects, you may refer to the "Configuring Properties During Object Creation" section of the MATLAB help. You can also find lots of useful information on numerous MATLAB forums on the Internet.

For example, if we want to connect to the www.myremotehost.com host (the fictional URL) at the port 9991 we might create a TCP object named as tobj using the following command:

```
tobj = tcpip('myremotehost.com', 9991);
```

After `tobj` has been created successfully it appears in the MATLAB workspace. Now we can operate with this TCP/IP object when implementing network operations. The common sequence when working with a TCP/IP object is as follows:

1. open the object using the `fopen()` function;
2. execute read/write operations using `fscanf/fread/fprintf` or another appropriate function;
3. close the TCP/IP object after completing operations using the `fclose()` function.

This is a very simplified algorithm, because we don't specify any additional parameters.

So, let's develop a distributed system where a MATLAB network client is used. This system will be built using a "client–server" approach; the TCP server will be written in C and the TCP client will be developed in MATLAB.
We take the DemoTCPServer application developed earlier as a TCP server (its source code is shown in Listing 5.15). The TCP server will receive some text string from the TCP client and display that string onto a screen. As to the client, we will design it "on the fly" by entering commands in the MATLAB command window (these commands may also be put in some M-file as well). Both server and client are running on the same machine, referred to as "`localhost`" in the TCP client and "127.0.0.1" in the TCP server. The server uses port 7171 for listening to incoming requests.
The C source code of the TCP server is given below (Listing 5.15):

Listing 5.15

```
#include <stdio.h>
#include <stdlib.h>
#include <signal.h>
#include <unistd.h>
#include <sys/socket.h>
#include <resolv.h>
#include <string.h>

#define port 7171

int server;
pid_t PID;
```

```
void sig_handler(int signal_num)
{
  close(server);
  printf("Listening socket has closed.\n");
  kill(PID, SIGTERM);
}

int main(void)
{
 int client;
 struct sockaddr_in addr;
 char buf[4096];
 int bWritten, bRead;

 struct sigaction sa;
 addr.sin_family = AF_INET;
 addr.sin_port = htons(port);
 addr.sin_addr.s_addr = inet_addr("127.0.0.1");

 server = socket(AF_INET, SOCK_STREAM, 0);
 bind(server, (struct sockaddr*)&addr, sizeof(addr));
 listen(server, 15);

 memset(&sa, 0, sizeof(sa));
 sa.sa_handler = &sig_handler;
 sigaction(SIGINT, &sa, NULL);
 PID = getpid();
 printf("TCP Demo Server waits for incoming requests on port 7171\
n");
 printf("Press Ctrl-C to terminate the program...\n");

 for(;;)
 {
  client = accept(server, 0 , 0);
  bRead = recv(client, buf, sizeof(buf), 0);
  if (bRead > 0)
  {
   buf[bRead] = '\0';
   printf("RECEIVED: %s\n", buf);
```

```
    }
    close(client);
  }
  close(server);
  return 0;
}
```

Our TCP server waits for incoming requests on port 7171 and after a connection has been established takes data from the client and outputs it to the terminal window. We already discussed the TCP server's source code earlier, so we will not repeat that here. The TCP client will be built "on the fly" in MATLAB. Note that before executing the TCP client the server must be launched. Below you can see the command sequence which creates a TCP/IP object, establishes the connection to the TCP server and then sends a string to it:

```
>> t = tcpip('localhost', 7171);
>> t

   TCPIP Object : TCPIP-localhost

   Communication Settings
      RemotePort:          7171
      RemoteHost:          localhost
      Terminator:          'LF'

   Communication State
      Status:              closed
      RecordStatus:        off

   Read/Write State
      TransferStatus:      idle
      BytesAvailable:      0
      ValuesReceived:      0
      ValuesSent:          0

>> fopen(t);
>> fprintf(t, 'Test Message from MATLAB.');
>> t

   TCPIP Object : TCPIP-localhost
```

```
    Communication Settings
       RemotePort:            7171
       RemoteHost:            localhost
       Terminator:            'LF'

    Communication State
       Status:                open
       RecordStatus:          off

    Read/Write State
       TransferStatus:        idle
       BytesAvailable:        0
       ValuesReceived:        0
       ValuesSent:            26

>> fclose(t);
>> delete(t);
>> clear t;
```

The very first command in this sequence creates the TCP/IP object named as `t` by calling the `tcpip()` function. This function takes the name of the host ("localhost", in our case) and a number of the remote port (7171). After that is completed we check the object just created by entering its name `t` without a semicolon:

```
>> t
```

Note that after its creation the TCP/IP object remains closed (see the `Status` property). Another property that should be considered is `Terminator` that determines the end of byte stream. Both the client and server must use the same `Terminator` to escape from error upon data transfer. In our case, the `Terminator` property is set to '`LF`' by default; we can leave this parameter unchanged. Many times we are interested to know the value of `BytesAvailable` that indicates the number of bytes currently available to be read from the input buffer. The value of `BytesAvailable` property is continuously updated as the input buffer is filled. Initially, after the `fopen()` function has been called this property is set to 0.

To use the `BytesAvailable` property value we must read data asynchronously (default mode). When reading data synchronously control is taken by the MATLAB command window, which returns only after the input buffer is empty.

Another useful property is `ValuesSent` that indicates the total number of values written to the network application. Sometimes we may need this value to know how many bytes were written after completing the last write operation. This value is updated after each successful write operation, and is set to 0 after the `fopen()` function has been executed. If the program writes the terminator at the end of data buffer, then `ValuesSent` reflects this value. When writing data asynchronously, it is worthwhile using the `BytesToOutput` property which returns the number of bytes currently in the output buffer.
If the TCP/IP object `t` was created successfully, we can open this object and write data to the server. Two following statements do that:

```
>> fopen(t);
>> fprintf(t, 'Test Message from MATLAB.');
```

Here the `fopen()` function opens the TCP/IP object `t` and `fprintf()` sends the "Test Message from MATLAB" string to the server. If the `fprintf()` function succeeded we should see this string in the terminal window where the server is running. It is worthwhile checking the status of the network communication after each operation by entering the name `t` without a semicolon:

```
>> t
```

As you can see, the `Status` property is now set to "open" – it means that `fopen()` has succeeded. The `ValuesSent` property is set to 26 – that is the full length of the text string sent by the preceding `fprintf()` function, including the terminator.
The following three statements terminate communication between the client and the server:

```
>> fclose(t);
>> delete(t);
>> clear t;
```

The `fclose()` function closes the working connection to the TCP server; `delete()` and `clear` functions wipe off the TCP/IP object `t`.
This was a detailed description of how to implement a network communication from the MATLAB command window. As we now know how this stuff works, we can put the above commands in an M–file except those being used to display the state of TCP/IP object `t`. We can call such a file to execute each time we want to communicate with the TCP server. The M-file implementing the TCP client will contain the following statements:

```
t = tcpip('localhost', 7171);
fopen(t);
fprintf(t, 'Test Message from MATLAB.');
fclose(t);
delete(t);
clear t;
```

This very simple TCP client will be used as a basis for development of other projects discussed below. Applying the TCP client written in MATLAB and some TCP server allows you to build complex distributed systems. Let's discuss some useful demo projects based on MATLAB network capabilities.

Writing byte to the parallel port

In this project we will write a byte to the data port (address 0x378) of the parallel port. The TCP client running in MATLAB sends a byte to the TCP server. The latter writes the byte received directly to the data port. In this way we can remotely control external circuits attached to the parallel port from the MATLAB environment. Let's see the C source code of the TCP server shown in Listing 5.16:

Listing 5.16

```
#include <stdio.h>
#include <stdlib.h>
#include <signal.h>
#include <unistd.h>
#include <sys/socket.h>
#include <resolv.h>
#include <string.h>
#include <sys/io.h>

#define port 9191
int server;
pid_t PID;
```

```
void sig_handler(int signal_num)
{
  close(server);
  printf("Listening socket has closed.\n");
  kill(PID, SIGTERM);
}

int main(void)
{
 int client;
 struct sockaddr_in addr;
 char bVal;
 int bRead;
 struct sigaction sa;

 iopl(3);
 addr.sin_family = AF_INET;
 addr.sin_port = htons(port);
 addr.sin_addr.s_addr = inet_addr("127.0.0.1");
 server = socket(AF_INET, SOCK_STREAM, 0);
 bind(server, (struct sockaddr*)&addr, sizeof(addr));
 listen(server, 15);
 memset(&sa, 0, sizeof(sa));
 sa.sa_handler = &sig_handler;
 sigaction(SIGINT, &sa, NULL);
 PID = getpid();
 printf("Writing a byte to the data port 0x378 via port 9191\n");
 printf("Press Ctrl-C to terminate the program...\n");

 for(;;)
 {
   client = accept(server, 0 , 0);
   bRead = recv(client, &bVal, sizeof(char), 0);
   if (bRead > 0)
    outb(bVal, 0x378);
   close(client);
 }
 close(server);
 return 0;
}
```

Our TCP server works with a listening port 9191 waiting for incoming requests from the TCP client started in MATLAB. To gain access to our hardware from the server we need to introduce two functions, `iopl()` and `outb()` for accessing address 0x378. Recall, that `iopl()` function takes a single parameter that determines a level of privileges (we set this parameter to 3). The `outb()` function will write the byte received from the TCP client to the data port.
The main job is accomplished by the following section of code:

```
client = accept(server, 0 , 0);
bRead = recv(client, &bVal, sizeof(char), 0);
if (bRead > 0)
 outb(bVal, 0x378);
close(client);
```

If the client request has been accepted the server creates a working connection (the `client` socket) and takes the byte sent by the TCP client using the `recv()` call. This byte will be temporarily held in the `bVal` variable before writing to the data port. In general, it is worthwhile to know if something was received, so we check the condition `bRead > 0`. If the data byte sent is available (`bRead > 0`) we can write its value to the data port using the `outb(bVal, 0x378)` statement. After that we close the working connection by the `close(client)` call.
The MATLAB command sequence takes just several statements shown in Listing 5.17

Listing 5.17

```
>> t = tcpip('localhost', 9191);
>> fopen(t);
>> fprintf(t, '%d', 1);
>> fclose(t);
>> delete(t);
>> clear t;
```

The above code writes 1 to the address 0x378 using the `fprintf()` function. As usual, we first create a TCP/IP object named `t` and open it with the `fopen()` function. After a write operation is complete we close the `t` object using a `fclose()` call. The last two statements delete the `t` object from the MATLAB environment. The above source code may be saved in a M-file for further use.

Reading hardware registers via MATLAB

Reading data from a hardware register resembles the writing operation, although it requires some additional steps on the client side. Suppose that the TCP client running in MATLAB sends the byte to the TCP server which, in turn, writes data received to the data port (0x378). The MATLAB command sequence which should send the data byte to the server appears as follows (let's assume that the TCP server listens to port 9191):

```
>> t = tcpip('localhost', 9191);
>> t.Terminator = 'CR/LF';
>> fopen(t);
>> fprintf(t, 'ReadPort');
>> data = fread(t, t.BytesAvailable);
>> fprintf(1, 'Data port state: 0x%X', data);
Data port state: 0xD>> fclose(t);
>> delete(t);
>> clear t;
```

After the TCP/IP object `t` has been created and opened, the `fprintf()` function writes the "`ReadPort`" command string to the server. For clarity we have developed the TCP server so that it doesn't parse the command received; we may put in any string in the `fprintf()` function. After issuing the command string to the server we read a byte sent back by the server and save it in the data variable (the `fread(t, t.BytesAvailable)` function). Then the `fprintf()` function that follows outputs this data to the screen.
As usual, we close the TCP/IP object `t` using the `fclose(t)` function after all operations have ended. The last two statements wipe off the `t` variable. We can also save this command sequence in some M-file, say, `ReadParPort.m` and call it for execution when needed. Before saving the source code we need to slightly modify it. Finally, the content of M-file appears as follows:

```
global data;
t = tcpip('localhost', 9191);
t.Terminator = 'CR/LF';
fopen(t);
fprintf(t, 'ReadPort');
while (t.BytesAvailable == 0)
end;
data = fread(t, t.BytesAvailable);
fprintf(1, 'Data port state: 0x%X', data);

delete(t);
clear t;
```

Note that a `while()` loop was introduced in the above source code. This statement is necessary because the server takes some finite time before responding to the client, so the `BytesAvailable` property may be 0 until some time has passed after issuing the "`ReadPort`" command to the server.

Below (Listing 5.18) you can see the source code of a TCP server that listens to port 9191, reads the data port (0x378) when requested and sends data back to the MATLAB TCP client:

Listing 5.18

```
#include <stdio.h>
#include <stdlib.h>
#include <signal.h>
#include <unistd.h>
#include <sys/socket.h>
#include <resolv.h>
#include <string.h>
#include <sys/io.h>

#define port 9191
int server;
pid_t PID;

void sig_handler(int signal_num)
{
  close(server);
  printf("Listening socket has closed.\n");
  kill(PID, SIGTERM);
}

int main(void)
{
 int client;
 struct sockaddr_in addr;
 char test = 0xD;
 char bVal;

 char buf[128];
 int bRead;

 struct sigaction sa;
 iopl(3);
```

```
addr.sin_family = AF_INET;
addr.sin_port = htons(port);
addr.sin_addr.s_addr = inet_addr("127.0.0.1");

server = socket(AF_INET, SOCK_STREAM, 0);
bind(server, (struct sockaddr*)&addr, sizeof(addr));
listen(server, 15);
memset(&sa, 0, sizeof(sa));
sa.sa_handler = &sig_handler;
sigaction(SIGINT, &sa, NULL);
PID = getpid();

printf("TCP Server reads the data port 0x378 via TCP port 9191\
n");
printf("Press Ctrl-C to terminate the program...\n");
for(;;)
  {
   client = accept(server, 0 , 0);
   bRead = recv(client, buf, sizeof(buf), 0);
   if (bRead > 0)
    {
     outb(test, 0x378);
     test++;
     bVal = inb(0x378);
     send(client, &bVal, sizeof(char), 0);
    }
   close(client);
  }
 close(server);
 return 0;
}
```

The source code of the server should already be familiar, so let's focus on only the section within the `for()` loop. Once a working connection to the client is established (the `accept()` function succeeded) the program reads the byte stream sent by the client. As we already know, the client data may be a "`ReadPort`" string or something else because the server application doesn't try to recognize the command sent by the client. It is only required that the client data should not be of zero length (an empty string) because the `bRead` value has to be greater than 0 to successfully complete the read operation.

To visualize a data flow I used the little trick described earlier in this chapter. The value of the `test` variable is written to the data port using the `outb()` function, and then the same value is read back in the `bVal` variable. The server sends `bVal` to the client by calling the `send()` function and increments the `test` variable. Next time the client reads the data port, it receives this incremented value from the server.
All this is illustrated below. After the TCP server has started, we execute the ReadParPort.m file see the result. The output should look like this:

```
 >> ReadParPort;
Data port state: 0xD>>
>> ReadParPort;
Data port state: 0xE>>
>> ReadParPort;
Data port state: 0xF>>
```

You can see that the each subsequent byte read by `ReadParPort` is incremented from its previous value.
The last two projects described here may be considered as templates for more complex designs dealing with various electronics circuits attached to the parallel or serial port. For example you may want to read the state of hardware registers periodically by introducing a timer.
Let's discuss a more complex project. This will be a distributed system where the MATLAB environment with data processing in MATLAB is using a development board attached to the serial port.

Distributed system with a micro-controller development board

In this project we use two PCs (client and server) and a standalone micro-controller development board with an 8052 microcontroller device. Though based on an 8052 development board, this system may be easily adapted for PIC-, ARM7/9- or another micro-controller kit equipped with a serial interface. The functional scheme of our experimental system is shown in Fig.5.16.

In this system the micro-controller development board collects data from some external circuit and passes information to a local PC via a RS-232 interface. The TCP client application running in MATLAB takes a data byte via the RS-232 and sends it to the remote PC where the TCP server is running. Specifically, our demo distributed system checks the state of 8 TTL-compatible signals arriving at port P1 of the 8052 microcontroller. For this distributed system we need to develop three programs:

1. the control program for the 8052 development board which must read the 8-bit port P1 and send the byte captured to the MATLAB;

2. the TCP client that functions in the MATLAB environment. The client takes a byte delivered via the serial interface `/dev/ttyS0` and then passes data to the TCP Server;
3. the TCP server that listens to port 7171 and captures data from the client.

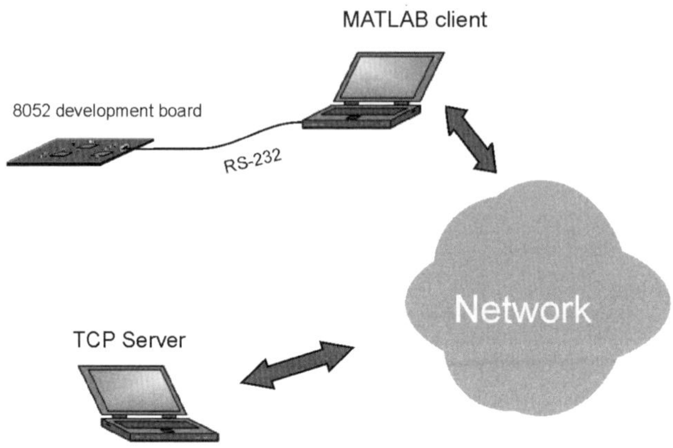

Fig.5.16

Let's begin from the program for the 8052 microcontroller board. This application is written in Keil C and its source code is shown in Listing 5.19:

Listing 5.19

```
#include <REG52.H>
#include <stdio.h>
#include <string.h>

void delay(void)
{
 long interval = 500;
 while(interval-- != 0);
}

int main(void)
{
```

```
char buf[128];
char cmd1[] = "ReadPort";
char cmd2[] = "*TEST?";
char data2[] = "8051 SYSTEM TESTED OK";

unsigned int pVal;
char i;
int args;

SCON = 0x50;
TMOD = 0x20;
TH1 = 0xF3;    //Baud rate 9600 at Fosc = 24.0MHz
PCON |= 0x80;
TR1 = 1;
TI = 1;

while (1)
{
 memset(buf, '\0', sizeof(buf));
 args = scanf("%s", buf);
 if (args == EOF)
  continue;
 i = strncmp (buf, cmd1, 8);
 if (i == 0)
 {
delay();
pVal = P1;
printf("State of P1: 0x%X\n",pVal);
  continue;
 }
 i = strncmp(buf, cmd2, 6);
 if (i == 0)
 {
delay();
printf("ANSW: %s\n",data2);
continue;
 }
 else
 {
```

```
  delay();
  printf("Wrong command\n");
  continue;
    }
   }
 }
```

The 8052 application reads the RS-232 interface and parses the string sent by the TCP client. There are two commands defined for our development board. After the "*TEST?" command (the cmd2 variable) has been identified the micro-controllers responds with the string data2. In the case of receiving the "ReadPort" command the 8052 program reads port P1 and sends the resulting byte back to the client. That's all about the 8052 software.

Let's discuss the TCP client. The client is realized in MATLAB and its source code is written to the ReadingInputs.m file (Listing 5.20):

Listing 5.20

```
% reading serial interface /dev/ttyS0

s = serial('/dev/ttyS0');
fopen(s);
fprintf(s, 'ReadPort');
cmd = fscanf(s);
data = fscanf(s);

% creating TCP client

t = tcpip('localhost', 7171);
fopen(t);

% sending data to the server

fprintf(t, '%s', data);
fclose(t);

% Closing objects

delete(t);
clear t;
```

```
fclose(s);
delete(s);
clear s;
```

First of all we need to create a serial port object named `s` using the following statement:

```
s = serial('/dev/ttyS0');
```

The following two statements open the serial port object and write the "`ReadPort`" command to it:

```
fopen(s);
fprintf(s, 'ReadPort');
```

The 8052 micro-controller must respond with the state of the `P1` port, an8–bit value. To read this byte the MATLAB program executes the following two statements:

```
cmd = fscanf(s);
data = fscanf(s);
```

The `cmd` variable will hold the command string, while the `data` variable contains the data byte sent by the 8052. The TCP client then sends `data` to the TCP server, but before doing that we must first create a TCP/IP object through which we send data bytes to the server. The following statement does this:

```
t = tcpip('localhost', 7171);
```

After executing the above statement the `t` variable will identify a TCP/IP object.
The `fopen(t)` statement that follows opens the working connection to the TCP server and the program sends data using the `fprintf(t, '%s', data)` statement. The `delete()` and `clear()` statements destroy both serial and TCP/IP objects.

The TCP server is implemented as a high-performance server based on pre-forking. This means, that the server creates a number of child processes while starting up, each of them services the separate client request. This allows handling of new clients with minimal delays and provides more performance for multiple clients. With such a TCP server your hardware may be handled simultaneously from several remote devices (PCs, mobile phones, etc.) without loss of performance. Pre-forking is widely used when designing high-performance server applications. For instance, the popular Web server Apache has been developed using this programming technique.

Our TCP server starts in the usual way, creates a listening socket on port 7171 and forks the predefined number of child processes (5 in our case). After the client establishes a connection, one of the running child processes services data transfer and exits. The next request will be serviced by another process and so on. The server tracks the number of child processes running; as some child process terminates, the server immediately creates a new one.
This was a brief description on how the pre-forking TCP server functions. The source code of a TCP server (DataServer1.c file) is shown below (Listing 5.21):

Listing 5.21

```
#include <stdio.h>
#include <stdlib.h>
#include <signal.h>
#include <unistd.h>
#include <sys/socket.h>
#include <resolv.h>
#include <string.h>

#define port 7171
#define MAXCON 5

int cntChild = 0;
pid_t pid;
void sig_child(int signal_num)
{
 wait(0);
 cntChild--;
}

int main(void)
{
 int server, client;
 struct sockaddr_in addr;
 char buf[4096];
 int bWritten, bRead;
 struct sigaction sa;

 addr.sin_family = AF_INET;
 addr.sin_port = htons(port);
 addr.sin_addr.s_addr = inet_addr("127.0.0.1");
```

```
 server = socket(AF_INET, SOCK_STREAM, 0);
 bind(server, (struct sockaddr*)&addr, sizeof(addr));
 listen(server, 15);
 memset(&sa, 0, sizeof(sa));

 sa.sa_handler = &sig_child;
 sigaction(SIGCHLD, &sa, NULL);
 printf("Data Server waits for requests on port 7171\n");
 printf("Press Ctrl-C to terminate the program...\n");

 for(;;)
 {
  if (cntChild < MAXCON)
  {
   if ((pid = fork()) == 0)
   {
    client = accept(server, 0 , 0);
    if (client > 0)
    {
     bRead = recv(client, buf, sizeof(buf), 0);
     if (bRead > 0)
     {
      buf[bRead] = '\0';
      printf("RECEIVED: %s\n", buf);
     }
    }
    close(client);
    close(server);
    exit(0);
   }
  else if (pid > 0)
   cntChild++;
  else
   printf("fork() failed.\n");
 }
 sleep(1);
}
close(server);
return 0;
}
```

As the source code of our pre-forked TCP server may seem a little complicated, we will analyze it in detail. This server may fork 5 child processes – this is defined by a `MAXCON` constant. The current number of processes being executed is held in the `cntChild` variables. This variable is decremented in the `sig_child` handler which is called while the child process terminates. This handler processes the `SIGCHILD` event that fires when a child process is complete. On the other hand, when a new process is created the `cntChild` variable will be incremented. Thus the TCP server provides 5 pre-forked processes running simultaneously.

The main job is implemented in a `for()` loop. First, the server checks the number of active processes running by executing an`if...else` control statement:

```
if (cntChild < MAXCON)
{
. . .
}
. . .
else
 sleep(1);
```

When there are no incoming requests from clients all 5 child processes are in an idle state. In this case the server gets paused for 1 s by invoking the `sleep(1)` function and checks the number of child processes again. Suppose, some request has arrived at the server. Since then some of the running processes (usually with the lowest PID) take care of servicing this request. Initially, a new working connection is created by the following statement:

```
client = accept(server, 0 , 0);
```

After the `client` socket has been successfully created (the `accept()` call returned a positive value), the `if` control statement that follows reads data through the open connection:

```
if (client > 0)
{
 bRead = recv(client, buf, sizeof(buf), 0);
 if (bRead > 0)
 {
  buf[bRead] = '\0';
  printf("RECEIVED: %s\n", buf);
 }
}
```

```
close(client);
close(server);
  exit(0);
```

The byte stream is held in the `buf` variable and then output using the `printf()` library function. The child process closes the active socket by executing the `close(client)` function. As the child process does not need a copy of the descriptor of the `server` socket, we can close `server` using the `close(server)` function. Recall that a child process when created gets all file descriptors of its parent process, so it is safe to close them upon exit. The last statement in this section is the `exit(0)` function that terminates the given child process.
Meanwhile the parent process increments the value of the `cntChild` variable in the `else` part of the `if...else` statement:

```
else if (pid > 0)
 cntChild++;
```

We must also consider the situation when the value of the `pid` variable is set to -1 after the `fork()` call has failed for some reason. In that case only the error message is displayed.
Since we need to decrement the `cntChild` variable after the child process has finished, signal handling is introduced in this program. In Linux, the termination of a child process may be captured by the `SIGCHLD` event handler. The `sig_child()` function that is called when a child process terminates, simply decrements the `cntChild` variable.

Below (Fig.5.17) you can see the terminal window of a running TCP server (the DataServer1 executable):

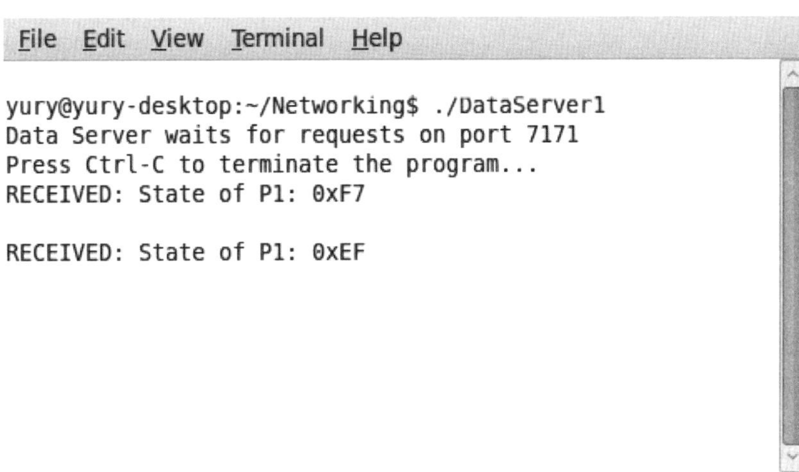

Fig.5.17

As you can see, the TCP server had processed two requests that came from the MATLB client. The first measurement indicated that the state of `P1` was 0xF7, while the second measurement yielded 0xEF. Both results were gained after executing `ReadingInputs.m` file twice (Fig.5.18):

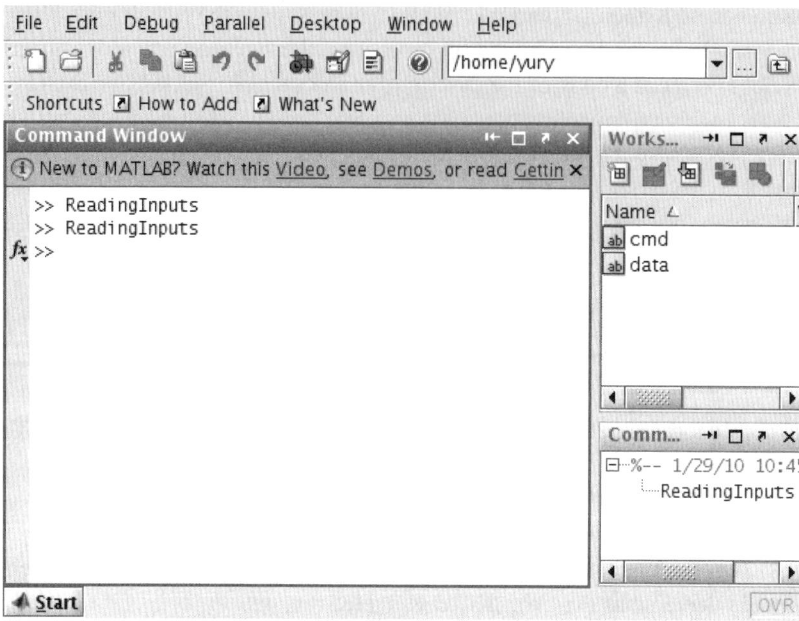

Fig.5.18

To experiment with this project the TCP server must be started first. We can see how our pre-forking server functions by entering the `ps -ef` command three times. This is illustrated in Fig.5.19:

After the TCP Data Server has started the first `ps -ef` command showed 6 processes including one parent and five child processes. Upon executing `ReadingInputs.m` script for the first time the child process with PID 2188 was executed and terminated; meanwhile the new process with PID 2196 has been created by the server. We can see that after executing the second `ps -ef` command.

Upon executing `ReadingInputs` once more the child process with PID 2189 terminated after data processing. Immediately the new child process with PID 2201 started to run. That may be seen after a third `ps -ef` command was executed. As you can see from Fig.5.19, the TCP server permanently maintained 5 active child processes. Note that our pre-forking TCP server is very simple – you may use it as a template for designing your own server application with advanced features.

With such a high-performance TCP server you can collect data from different clients scattered over a network. You can have, for example, some MATLAB clients reading input signals from various devices (microcontrollers, parallel or serial interfaces, etc.) and sending data gathered to your TCP server. In this case we need to somehow identify the source that issued the request. The most common method is based on reading the client's IP-address which explicitly identifies the source. Nevertheless, you may apply your own method for identifying the network client. For example, you may put additional unique information in a message which a particular client sends to a server. In this case the TCP server must parse the message and define the source of information just received.

File Edit View Terminal Help

```
yury@yury-desktop:~$ ps -ef | grep DataServer1
yury      2187  2102  0 15:39 pts/1    00:00:00 ./DataServer1
yury      2188  2187  0 15:39 pts/1    00:00:00 ./DataServer1
yury      2189  2187  0 15:39 pts/1    00:00:00 ./DataServer1
yury      2190  2187  0 15:39 pts/1    00:00:00 ./DataServer1
yury      2191  2187  0 15:39 pts/1    00:00:00 ./DataServer1
yury      2192  2187  0 15:39 pts/1    00:00:00 ./DataServer1
yury      2194  2147  0 15:39 pts/2    00:00:00 grep --color=auto DataServer1
yury@yury-desktop:~$ ps -ef | grep DataServer1
yury      2187  2102  0 15:39 pts/1    00:00:00 ./DataServer1
yury      2189  2187  0 15:39 pts/1    00:00:00 ./DataServer1
yury      2190  2187  0 15:39 pts/1    00:00:00 ./DataServer1
yury      2191  2187  0 15:39 pts/1    00:00:00 ./DataServer1
yury      2192  2187  0 15:39 pts/1    00:00:00 ./DataServer1
yury      2196  2187  0 15:40 pts/1    00:00:00 ./DataServer1
yury      2198  2147  0 15:40 pts/2    00:00:00 grep --color=auto DataServer1
yury@yury-desktop:~$ ps -ef | grep DataServer1
yury      2187  2102  0 15:39 pts/1    00:00:00 ./DataServer1
yury      2190  2187  0 15:39 pts/1    00:00:00 ./DataServer1
yury      2191  2187  0 15:39 pts/1    00:00:00 ./DataServer1
yury      2192  2187  0 15:39 pts/1    00:00:00 ./DataServer1
yury      2196  2187  0 15:40 pts/1    00:00:00 ./DataServer1
yury      2201  2187  0 15:42 pts/1    00:00:00 ./DataServer1
yury      2203  2147  0 15:42 pts/2    00:00:00 grep --color=auto DataServer1
yury@yury-desktop:~$
```

Fig.5.19

Home wireless systems

In the previous section we discussed distributed systems based on using some popular network protocols. Nevertheless, there is another way to design distributed systems – by using wireless technologies. It may at first seem too complicated to design such a sort of distributed system, especially for inexperienced readers, but, luckily for us we can apply already proven and ready-to-use devices which we slightly modify for our needs.

First, let's imagine how the simplest wireless systems work. Everywhere in this section, when discussing wireless systems we will mean those operating at radio frequencies (RF) of about 300 MHz or higher. Usually the simplest wireless system provides a unidirectional data transfer from an RF transmitter to an RF receiver, although there are many systems capable of implementing a bidirectional data flow. To pass the useful signal using RF we must apply some kind of modulation that allows placing our data into a RF carrier frequency. The most popular type of modulation used in relevant wireless devices is "Frequency Modulation" (FM) which changes the carrier frequency over a narrow band according to changes in amplitude of the modulating signal. Many times you can see the abbreviation "FM transmitter" or "FM receiver" indicating that FM modulation is applied to the particular transmitter/receiver.

As a basis for building our wireless projects – and to show how simple it can be -we will use a low-cost, ready-to-use wireless doorbell. That device may be easily found almost everywhere on the marketplace. The typical wireless doorbell consists of two modules, an FM transmitter and FM receiver. The FM transmitter module passes the signal each time when a user presses the call button on its front panel. At this moment power supply gets connected to the RF oscillator which beams the signal modulated by a lower frequency, usually 32 KHz, over a distance of about 30 – 50m. This signal may be captured by the FM receiver module located inside that distance from the transmitter. Once the signal from the transmitter has been detected, the receiver yields a sound signal or some melody going to a built-in speaker.

So, how could we adapt our wireless doorbell to operate as a control system handled by a PC's software? There are several ways to achieve that. We might replace, for example, the mechanical button in the FM transmitter with an electronic switch controlled by software running on the PC – this allows switching on/off the load (relay, motor, valve, etc.) attached to the FM receiver. It is also possible to attach some sensor connected via some conditioning circuit (a comparator, for example) to the FM transmitter. The conditioning circuit may control the FM transmitter, so the program running on the remote PC will know the state of some physical parameter (temperature, pressure, etc.).

In spite of a variety of wireless doorbells most of them actually operate in a similar manner. Each device is activated by pressing a user button which connects the power supply (9 – 12 V battery) to the FM transmitter circuit. Being pressed for a short time the button causes a radio signal transmission to the FM receiver. Upon receiving the FM signal the receiver enables a sound synthesizer chip to provide some sound signal or melody that goes to a speaker. When released, the button cuts off the power feeding the FM transmitter, so the transmission ends. Connecting the mechanical switch to the power circuit of the transmitter prevents current from flowing when not in use and thereby prolongs the battery life. We can control the FM transmitter electronically by replacing the mechanical button with some electronic circuit being handled by software.

Another common feature that can be applied to home projects arises from the FM receiver circuit. As was mentioned above, upon receiving a signal from the FM transmitter the receiver enables a sound synthesizer chip which generates some sound. The sound chip provides a relatively powerful signal that can directly drive a low-impedance load such as a speaker. We may exploit this signal and redirect it to our control circuit that drives some load. In this way we can use the sound signal to turn on/off actuators, relays, valves, etc.

So, let's go to the practice. Our first project will be a simple but effective wireless security system with a Passive Infrared Sensor (PIR-detector or PIR-sensor). Recall, that we have already built a simple security system using the PIR-detector circuit attached to the serial interface. Now we want to design a wireless variant of the security system with this same sensor.

As you will recall, the PIR-detector tracks a (human body heat) movement and issues a signal which may be further processed in some appropriate way. In our system the signal generated by the PIR-detector switches on the FM transmitter which, in turn, beams the FM signal to the remote FM receiver. There the sound chip is enabled and the sound output signal passes to a circuit that converts sound into a TTL-compatible signal. This signal arrives at an input of the parallel interface which is monitored by software. The program checks to see if the voltage level of the signal has dropped.

When this occurs, the program interprets it as an alarm signal and outputs the appropriate message to a screen. This system may be useful when guarding your home or some outbuilding located somewhere nearby.
To make things clearer, let's see the simplified functional diagram of our wireless security system shown in Fig.5.20:

As you can see, both the PIR–detector and FM transmitter may be placed, for example, within some remote building we want to protect, while the FM receiver may be placed away from this by some distance within range that it would be possible to reliably receive the alarm signal. You must remember that the radio signal of a very high frequency may be significantly attenuated or even lost when it is passed through some physical barrier (e.g. building, metal fence, etc.), so it is worthwhile taking time to place both transmitter and receiver so to ensure a good r.f. connection.

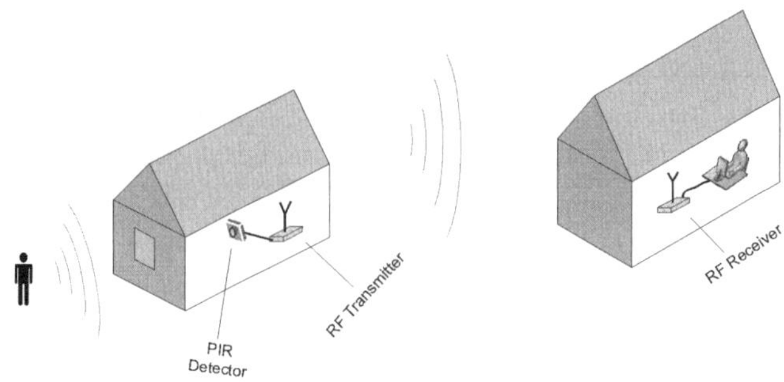

Fig.5.20

Note that the distance of a reliable communication between the transmitter and receiver depends on both the power of the signal emitted by the transmitter and the sensitivity of the receiver. With this particular system the distance may vary from about 30 to 50 m. If you are experienced in RF systems, you may increase the distance by introducing an RF power amplifier driven from the RF oscillator stage of the transmitter. Another way to extend the range of action of the transmitter is to apply an external antenna (piece of wire) about 10 – 15 cm long. The antenna is added to the point where the LC tank of the RF oscillator is connected to the collector of the transistor. To lessen the loading influence of the antenna on the stability of the carrier frequency it would be worth connecting the antenna to the LC tank via a capacitor of about 10 – 20 pF. Of these two options, use of an RF amplifier would be the technically preferred solution allows isolation of the RF oscillator from incoming interference and increasing power of the output RF signal. You may combine both approaches mentioned to obtain an even better performance. Anyway, I encourage you to explore the numerous technical sources concerning design of RF circuits.

After the signal has been captured it will be converted to a TTL-compatible voltage level passing to an input of the PC's parallel interface. Once a voltage drop is detected (the PIR-sensor has fired) the control program outputs an appropriate message onto a screen. This is in general how the wireless security system will function.

To build our system, therefore, the following components were taken:

- PIR-detector (PIR-sensor). A Passive Infrared Sensor is an advanced version of a motion sensor. PIR sensors 'know' how to identify people using the infrared radiation that people emit, which stems from body heat. The PIR sensor is completely passive and reacts only to drastic changes in levels of infrared radiation emitted in the surveillance area, caused by the movement of a person in a room. This prevents false alarms when the floor cools at night, for instance. These sensors are so designed that they don't react to pets moving around in their surveillance area. This part may usually be purchased from some of the larger electronic shops through the Internet. When seeking out a PIR-sensor select a fully assembled module that includes the PIR-sensor itself, conditioning circuit and Fresnel lens to provide the greatest distance of the protected area. Usually such a module provides a voltage swing in the range of about 3 to 12 volts. In my particular case I took a module offering a low-to-high voltage swing from 4 through 12 V upon an intrusion being detected.

- FM transmitter and FM receiver. I took a very low-cost (about $8) wireless doorbell which consisted of both FM transmitter and FM receiver configured for operation at ~433 MHz. (Note that these devices usually operate at ~315 -433 MHz). As the power output of the radio signal is limited, the reliable connection reached up to 50 meters.

So, let's discuss the system in detail. After you have all the components necessary you may start to assemble your system. In our case we needed to modify both the FM transmitter and receiver. First of all we connected our PIR-sensor to the FM transmitter. The PIR-detector when fired (ie. detector activated) raised the voltage level on its output; the high level held constant for some time, usually 0.5 – 1 s, that was sufficient to power up the FM transmitter and send an alarm signal. As the signal on the PIR-sensor's output is low-power, we applied an electronic switch that powered up the FM transmitter. This idea was realized in the following schematic circuit shown in Fig.5.21:

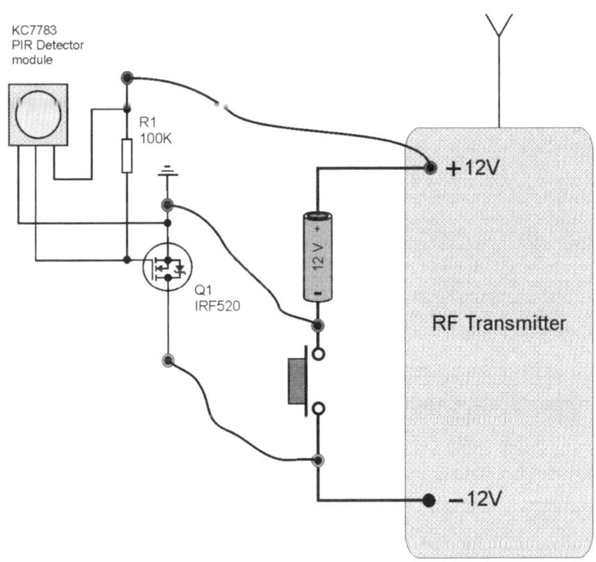

Fig.5.21

The PIR detector module type KC7783 was used again, described earlier. This is a fully assembled and tested miniature PIR module with only 3 connections: Power, Ground and Output. The module can be powered from +4 – +12V DC power supply (in our project we used the +5V power source). The output signal is an active high pulse of approximately 0.5 seconds and remains active as long as there is motion. The device can "feel" the motion over within a sensitive range up to 10 feet. The size of the module is 35mm x 25mm so it can be easily mounted on a wall. In fact, you can freely apply almost any manufacturers PIR-sensor, for they pretty-much all generate TTL-compatible signals. After the PIR–sensor has been activated it raises its output voltage. The high level voltage opens the power MOSFET IRF520 that supplies the power to the RF transmitter. As you can see, the IRF520 is connected in parallel with the button, so we can turn on/off the transmitter circuit electronically also. The power MOSFET IRF520 may be substituted with any similar device, such as IRF510, IRF530, etc.

The FM receiver must be changed as well. We can take the audio signal at the speaker input and convert this signal into a TTL-compatible signal. The appropriate circuit is shown in Fig.5.22:

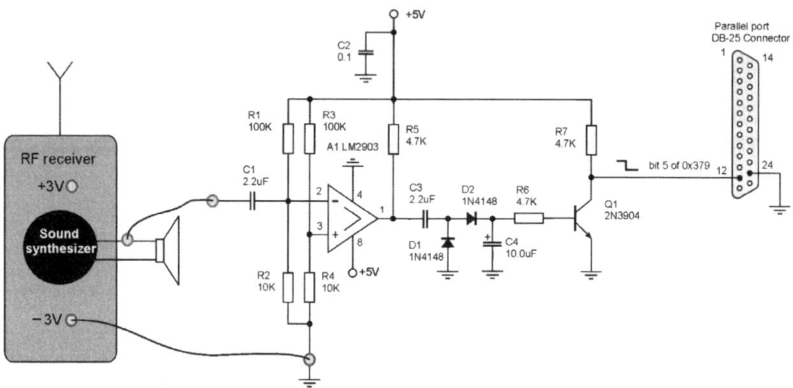

Fig.5.22

To build this circuit you need to find the sound output in your FM receiver module. I mean the wire that goes from the sound synthesizer chip to the speaker. Usually a standard FM receiver module of the wireless doorbell produces some melody stored in a flash memory of a sound synthesizer chip.

The power of the signal on the sound synthesizer output is enough to drive a low-impedance load like the speaker. In our case we take the signal from the sound chip and convert it to a TTL-compatible signal that appears as a voltage swing on the collector of the Q1 switch. The Q1 output, in turn, is read by software through bit 5 of the status port (address 0x379).

As a signal converter we used a simple comparator circuit based upon a LM2903 chip. The signal from the sound chip passes to the inverting input of A1 through the capacitor C1. The output signal on pin 1 of the LM2903 feeds the rectifier (diodes D1, D2 and capacitor C4). The comparator produces a pulse train on pin 1 representing when the audio signal is present on the sound chip output.

This sequence feeds the rectifier which, in turn, feeds a DC voltage into the base of Q1. The Q1 transistor opens and the voltage on its collector drops letting the program know that the PIR-sensor has fired. Otherwise, if the sound signal is absent or at about 0 V the collector voltage of Q1 remains high as Q1 is off. Bit 5 also would then be high. The program keeps polling this bit in an endless loop.

Note that both circuits shown in Fig.5.21–5.22 illustrate the common approach that may be applied to any wireless doorbell. Anyway you need to carefully examine your particular devices (FM transmitter and FM receiver) before embarking on any action. When inspecting your transmitter module try to find the traces going to the mechanical button – this will help you to build the circuit shown in Fig.5.21. To use the circuit from Fig.5.22 you must find the speaker inside your FM receiver module, then detach the signal connection wire from the speaker and connect this wire to your converter as is shown in Fig.5.22.
The C source code of the relevant polling program is shown in Listing 5.22:

Listing 5.22

```
#include <stdio.h>
#include <sys/io.h>
#include <unistd.h>

int main(void)
{
 int bRead;
 iopl(3);
 while(1)
 {
  bRead = inb(0x379) & 0x20; // Reading bit 5 of a status port
  if (bRead != 0)
    continue;
  else
    printf("Alarm system was fired!\n");
  sleep(5);
 }
 return 0;
}
```

After gaining permissions to access hardware (function `iopl(3)`) the program enters the endless `while()` loop where bit 5 of the status port is constantly checked. At the beginning of each iteration the program code reads the status port and saves bit 5 in the `bRead` variable:

```
bRead = inb(0x379) & 0x20;
```

The `if` statement that follows checks the value of `bRead`. When `bRead` is non-zero it means that no signal has arrived from the PIR-sensor, so the program pauses for 5 s and the loop repeats. If the PIR-detector is activated, bit 5 is cleared so the program executes the `else` branch of the `if` statement where the alarm message is displayed. This demo program may be easily improved to implement more complicated tasks as reaction to the alarm signal.
Below (Fig.5.23) you can see the actual assembled PIR-detector circuit connected to the FM transmitter module.

Fig.5.23

Single-channel wireless control system

The FM transmitter/receiver pair may also be used to implement a computerized control over some remote device or actuator. In this case we can modify the FM transmitter as is shown in Fig.5.24:

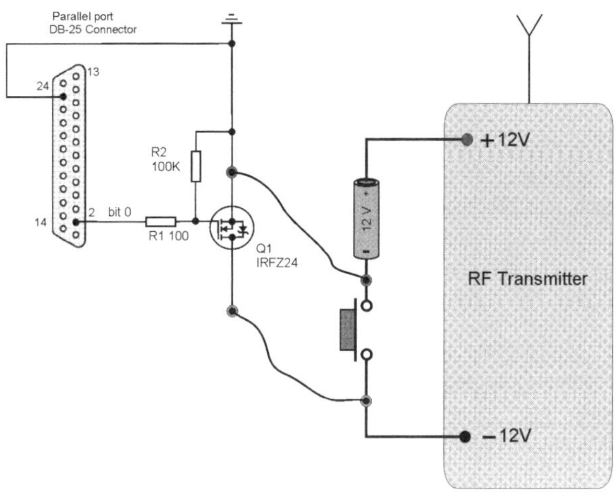

Fig.5.24

Here we connect bit 0 of the data port (address 0x378) to the gate of the power MOSFET IRFZ24 (Q1). Setting/clearing of this bit allows switching on/off of the RF transmitter. When the program sets bit 0, the switch Q1 is turned on thus feeding the transmitter. The FM receiver then picks up the signal that may be used to trigger some load. Thus we get the simplest one-channel control system. On the receiver side the FM signal may trigger, for example, a D-type flip-flop that, in turn, may switch on/off some power switch where a bipolar transistor or power MOSFET might be used. To control the load we can apply the circuit shown in Fig.5.22. In this case the signal on the collector of Q1 can drive, for example, a Darlington transistor or a power MOSFET.
The source code of the program which produces pulses that periodically turn on/off the FM transmitter is shown in Listing 5.23:

Listing 5.23

```
#include <unistd.h>
#include <stdio.h>
#include <sys/io.h>
```

```
#define dataPort 0x378

void main(void)
{

 char bVal = 0x0;

 iopl(3);
 while (1)
 {
  bVal = 0x1;
  outb(bVal, dataPort);
  sleep(1);
  bVal = 0x0;
  outb(bVal, dataPort);
  sleep(7);
 }
}
```

The source code is extremely simple. The program periodically sets/clears bit 0 of the data port in the `while()` loop thus providing short radio pulses issued by the FM transmitter.
You may also build bidirectional wireless systems by using two wireless doorbells. Note: In this case both devices must operate at different frequencies to avoid false triggering.

Simple remote control system using FM transmitter and receiver

We can also build simple wireless control systems using another approach where an FM transmitter and receiver are used. In this case we can take any low-cost FM receiver that will receive control signals from a low-power FM transmitter. In this configuration a control signal is fed to the FM transmitter from a sound card line output. The general scheme of such a wireless control system is shown in Fig.5.25:

How could we control some remote load attached to the FM receiver? Let's think of that. Suppose we want to remotely trigger a DC motor. The obvious way to do that is to pass the control signal wirelessly to the FM receiver, then take the audio signal at the output of the FM receiver, convert that signal into a digital form (for example, a TTL-compatible swing) and use that digital signal to switch on/off our load. Obviously, both the transmitter and the receiver must operate at the same carrier frequency.

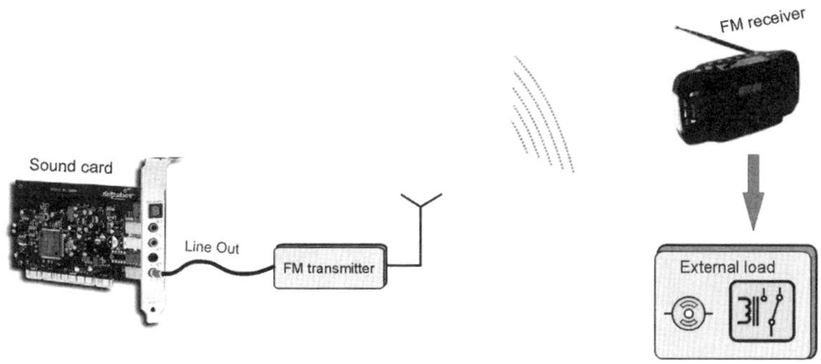

Fig.5.25

Generally, our wireless control system may look as shown in Fig.5.26:

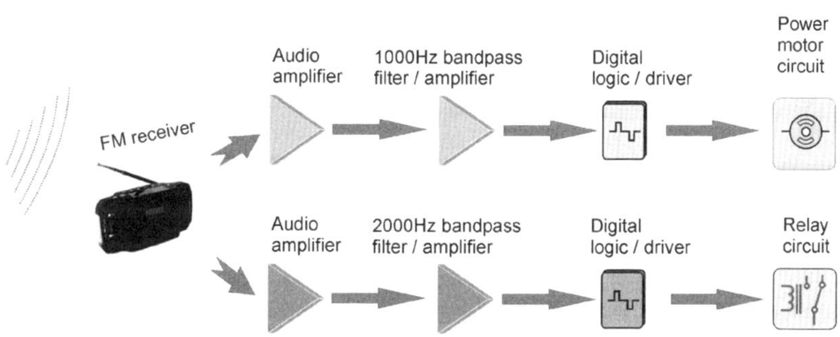

Fig.5.26

As you can see, we have created a scheme for a 2-channel system where a signal of 1000Hz is used for driving the DC motor, while that of 2000Hz triggers the relay. Thus, in using different frequencies we can easily expand our system, so as to control numerous loads. Let's discuss in detail one channel of such a system.

The transmitter part of our system may consist of a sound card connected to an FM transmitter. The control signal of the predetermined frequency created by audio software is delivered to the line output of the sound card. The FM transmitter input is fed from this signal that alters the carrier frequency of the transmitter. Thus, the FM carrier modulated by the audio signal travels to the receiver part of our system. As you already know, we can easily generate audio frequency signals directly in our application, or we can use popular software tools (Matlab/Simulink, LabView or some other popular audio editor, such as Audacity, etc).

Let's see what happens on the receiver part of our system. Once captured by the receiver, the signal is then demodulated, so we can hear a "pure" audio signal of 1000 or 2000Hz from the speaker. As we don't need acoustics, we must convert the given signal so that it can drive our motor or relay. The simple way to do that is to pick the signal out after the audio amplifier stage of the receiver, but prior to the speaker and redirect it to a bandpass filter. The bandpass filter, in turn, has to pass only signals whose frequency falls into its pass band and attenuate signals beyond the pass band.

As we use 1000Hz and 2000Hz signals, we must apply two bandpass filters at the output of the audio amplifier stage of the FM receiver. Depending on where the audio signal is taken, it might be necessary to amplify it before delivering to the control circuit. In that case we can introduce an amplification stage before the bandpass filter or we can provide the signal gain within the bandpass filter itself.

As we have formed an analog signal, we must convert it into digital that will switch on/off our load. Additionally, if our load consumes a high power, we might need an additional buffer driving such a load.

Let's go to the practice and see how our RC system may be implemented in "real life". We will design a single control channel operating at a frequency of 1000Hz – you can easily recalculate the system for operation with other frequencies. We will begin our discussion with the receiver part. First of all, we need to grab the audio signal obtained by the receiver. The best point to do that is the output of an audio amplifier of FM receiver. This is where the signal goes straight to the speaker, usually through a capacitor as shown in Fig.5.27:

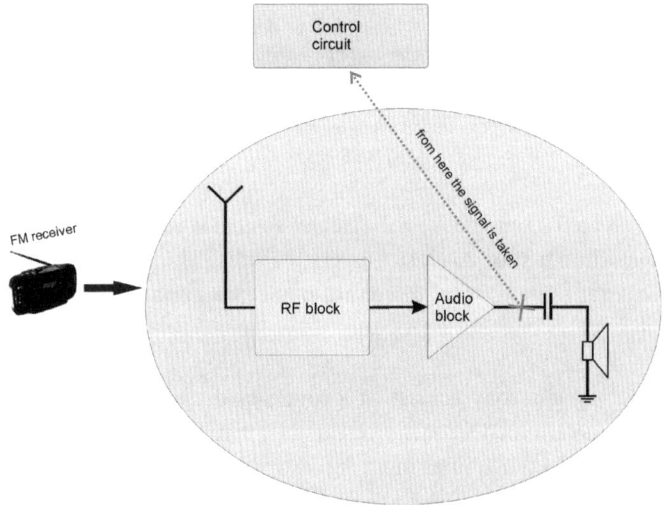

Fig.5.27

As you can see, the point where the signal for our control circuit may be taken from is between the output of the audio amplifier and the capacitor (usually electrolytic is used) through which the signal is coupled to the speaker. Anyway, if you didn't find this point within your FM receiver, you would need to probe the circuit and take any point where the audio signal emerges. If you have some experience in assembling electronic circuits, this task will not get you into trouble.

The next step is to amplify (if necessary) the audio signal just found, then to convert it into a digital representation. Finally, our digital signal will drive the load through some power switch. The demo control circuit which performs all that is shown below (Fig.5.28):

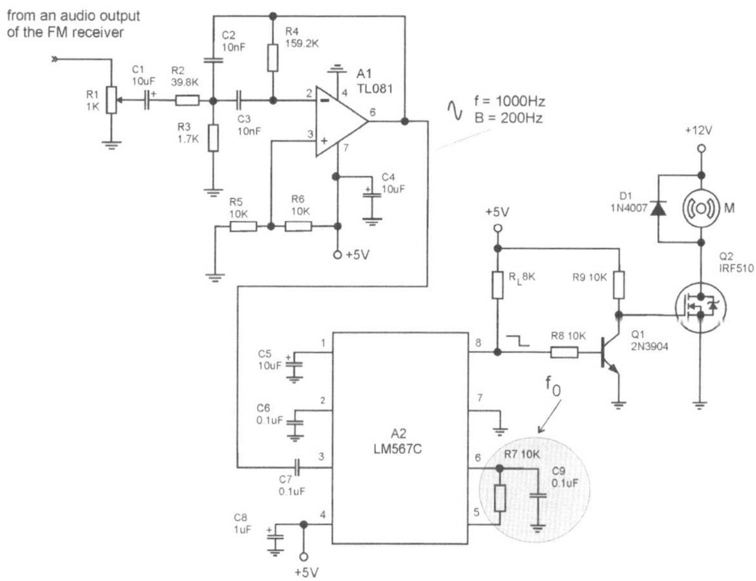

$$f_0 = \frac{1}{1.1 \times R7 \times C9} \qquad BW = 1070 \times \sqrt{\frac{V_i}{f_0 \times C6}} \text{ in \% of } f_0$$

Fig.5.28

As you can see, the audio signal is fed to the active filter input (just to the left side of R2 in the schematic) through a voltage divider R1 and a capacitor C1. You will have to adjust the level of the audio signal from the FM receiver so that it will be about 0.1 – 0.3V. If you pick up the signal, for example, after the last amplifier stage of the receiver, you surely will need to attenuate the signal, so that is where the adjustable divider may come in handy.

The active bandpass filter built around the TL081 opamp operates at a central frequency of 1000Hz with a bandwidth of 200Hz. Additionally, the filter provides a gain of 2V/V, so the magnitude of the signal on pin 6 of A1 is twice as high as that on the left lead of resistor R2.

Thus, input frequencies ranging from about 900 to 1100Hz will be amplified with a gain of 2 while signals beyond the bandpass will be lowered (attenuated).

The second stage that follows is a tone decoder where the popular chip LM567 is used. This stage is configured to trigger its output when frequencies around 1000Hz arrive at pin 3. As you already know, the LM567 chip recognizes the signal within the configured passband resulting in a voltage drop on pin 8. This level is kept low until the frequency of the input signal goes outside the given range. Signals outside the passband draw pin 8 high. The signal on pin 8, in turn, drives the power switch (Q1 and Q2) that connects/disconnects the power to the motor M. The low level on pin 8 opens Q2 thus feeding power to the motor M, while the high level breaks the power to the motor.

In the same way we can configure the control circuit to function with signals around 2000Hz or another frequency. For each particular frequency we need to recalculate appropriate values of time dependent components in both the filter and the tone decoder. This circuit is not something perfect; can easily improve it as you want.
That's all about the receiver part of our control system. Let's go to the transmitter which delivers the control signal from the PC to our receiver. The transmitter circuit is extremely simple and includes only a few components. It is designed as a Colpitts oscillator connected in a common base configuration relating to RF signals (Fig.5.29):

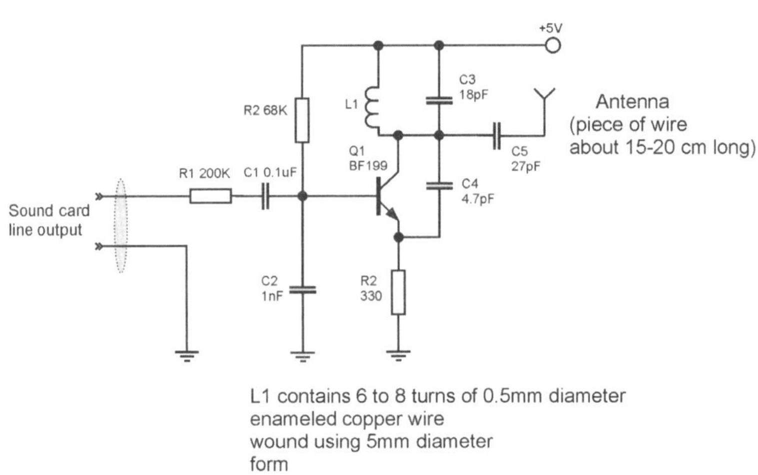

Fig.5.29. Simple FM transmitter circuit

The C2 capacitor of 1nF ensures that the low-frequency audio signal will freely travel to the base of Q1 while RF frequencies go straight to the "ground". The frequency range of the transmitter is defined by the resonant circuit (tank) composed of L1, C3 – C4 components and the base – emitter capacitance of the transistor Q1 (BF199). The audio signal voltage swing when applied to the base of Q1 changes the inner capacitances of the transistor thus providing an angle (FM) modulation of a carrier frequency of the oscillator. With the given values the transmitter operates in a center range of about 88 to 108 MHz; the central frequency is adjusted by mechanically squeezing or stretching L1, no another tuning is needed. You should adjust the coil until the FM radio goes silent – this indicates that the carrier frequency from the transmitter is being picked up by the receiver. The signal of the transmitter may be received within 100m around or so. The drawing below is the copper side of the pcb for this transmitter which I assembled for this project (Fig.5.30):

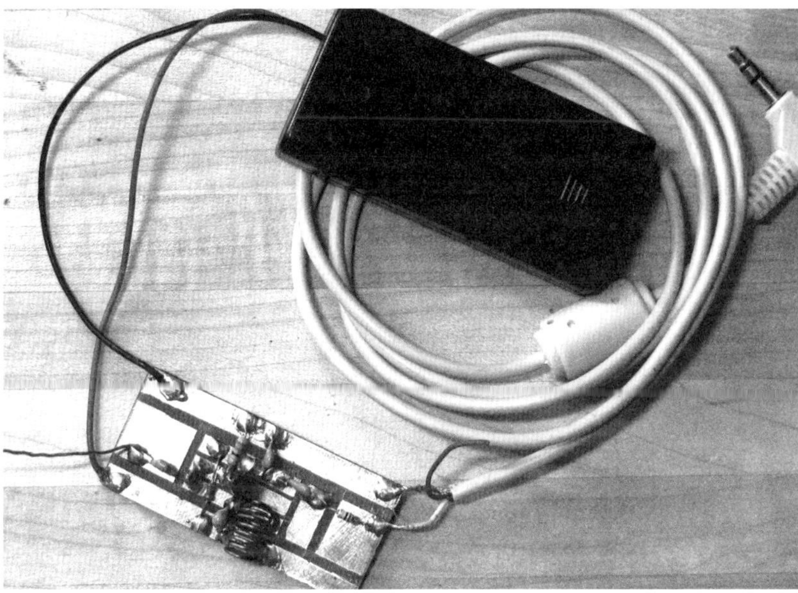

Fig.5.30. Assembled FM transmitter

Some words about the above design. You can apply almost any high-frequency bipolar transistor (2N3904, 2N4401, 2SC9018, etc.). The circuit can operate over a wide range of supply voltages (I tested it at 2.7V through 5V). Remember that when you change the supply voltage the inner capacitances of Q1 also changes and that results in altering of the carrier frequency of the oscillator, so you will have to re-adjust the resonant circuit as mentioned above.

To extend a distance you can either increase the sensitivity of your receiver (for example, by adding an RF amplifier stage before the receiver input) or increase the power of the signal from your transmitter. Since designing and tuning the small signal RF amplifier for the receiver may be too difficult for inexperienced readers, I would like to offer a modified variant of the FM transmitter just discussed. I simply added a power RF amplifier to the output of the RF oscillator, so it resulted in a higher power output signal. The modified FM transmitter circuit appears as follows (Fig.5.31):

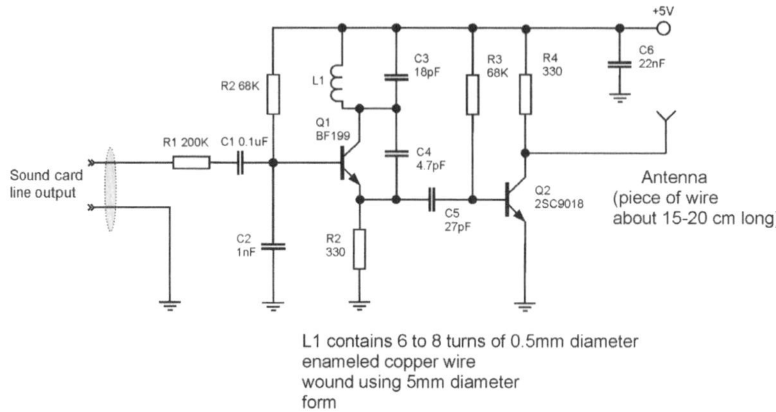

Fig.5.31. Advanced FM transmitter

The RF amplifier circuit uses the Q2 transistor that may be the same as that of the RF oscillator or you can use almost any RF bipolar transistor instead (I used type 2SC9018, although BFS17, BFR92, BFR93A will suit well). The RF amplifier is connected as a common emitter stage and is biased through the R3 resistor; the RF signal from the oscillator is fed to the base of Q2 through the capacitor C5. The amplified RF signal from the collector of Q2 goes straight to the antenna. As we have no resonant circuits here, this stage doesn't require any tuning. The great advantage of this modified circuit is that we get a higher stability of the carrier frequency because the RF amplifier isolates the oscillator circuit from the antenna. This circuit may also operate from about 2.7 through 5V, but you will need to adjust the resonant circuit of the oscillator stage after you have altered the supply voltage to the transmitter. Additionally, you can extend the distance of a reliable transmission if you apply the longer (30 – 40 cm) antenna.

Conclusion

We have considered designs of distributed systems using the network capabilities of the Linux operating system and MATLAB. In most projects we applied a TCP/IP protocol stack as a basis for communication between applications. The high reliability and versatility of the TCP networking allows us to easily and confidently connect various systems and applications. With TCP/IP you can easily convert your local system into a distributed system consisting of different subsystems developed with different hardware/software tools. Due to its inter-processing capabilities the Linux operating system allows building of distributed systems running on the same computer – this comes in handy when different parts of such systems are written in different languages. The same is also true for TCP/IP networked systems that may be successfully executed on the same host. The home and laboratory distributed systems controlled by Linux may also be built using low-cost wireless devices with minimal modifications.

Index

A

B

C

D

E

F

G

H

I

K

L

M

N

O

P

R

S

T

U

V

W